BrainBriefs

BrainBriefs

ANSWERS TO THE MOST (AND LEAST)
PRESSING QUESTIONS ABOUT YOUR MIND

Art Markman, PhD and Bob Duke, PhD
Hosts of the *Two Guys on Your Head* Podcast

STERLING
New York

STERLING
New York

An Imprint of Sterling Publishing, Co., Inc.
1166 Avenue of the Americas
New York, NY 10016

ISBN 978-1-4549-1907-0

Distributed in Canada by Sterling Publishing Co., Inc.
c/o Canadian Manda Group, 664 Annette Street
Toronto, Ontario, Canada M6S 2C8
Distributed in the United Kingdom by GMC Distribution Services
Castle Place, 166 High Street, Lewes, East Sussex, England BN7 1XU
Distributed in Australia by NewSouth Books
45 Beach Street, Coogee, NSW 2034, Australia

For information about custom editions, special sales, and premium and corporate purchases, please contact Sterling Special Sales at 800-805-5489 or specialsales@sterlingpublishing.com.

Manufactured in the United States of America

4 6 8 10 9 7 5 3

www.sterlingpublishing.com

Design by Philip Buchanan

To Rebecca, without whom there would be no Guys.

Contents

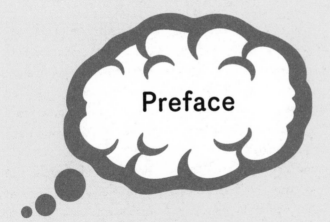

Preface

WHEN WE FIRST STARTED THINKING ABOUT WRITING A BOOK RELATED to our radio show/podcast *Two Guys on Your Head*, it seemed like a daunting task to try to capture the spirit of a conversation in a book. On the one hand, the format of the show (as a seven-and-a-half-minute segment) is a natural fit with short chapters on different topics. On the other hand, while the two of us sound alike, we have very different (complementary) outlooks, insights, and dispositions. (Plus, it is not always easy to be funny in print—the timing and intonation that help to sell a joke are often missing on the page.)

Fortunately, we think we've captured, in book form, the essence of our personalities and perspectives as we tackle an array of interesting, head-scratching quandaries about human behavior. But first, we want to say a few words about the arrangement to get you oriented.

Like the episodes of the show, the book is in no particular order. You could choose to read it straight through from beginning to end. Or you may want to skip around. While we have a lot of strong opinions—we're college professors, after all—we endeavored to make each conversation as deeply research-based as possible. At the end of the book, you'll find a list of publications that inspired our discussions for each chapter. Those readings are a great place to start for people interested in further study of various topics in the field of psychology.

We know that we have not done justice to all of the subtlety in the areas we write about. This book is meant to give you a flavor of the latest research occurring across a broad spectrum of the field. We hope that, at minimum, you will find the book to be a source of entertainment, one that piques your curiosity, prompts conversation with friends, and perhaps provides a few nuggets of insight. Even better if you can apply what you learn to your own life or find yourself wanting to dig deeper into the study of human nature.

And with that, we invite you to dive in. Welcome to what we both believe is the vastly interesting world of the human mind, and to *Two Guys on Your Head*. We're glad you're here.

Does being open to experience lead to success?

ACADEMICS LOVE TO TALK TO OTHER ACADEMICS. WE MEET AT CONferences and share our research, reveling in conversations that we're pretty sure only twenty-five other people on the face of the planet would care about. We are with "our people." Of course, we also gossip about other academics and complain about aspects of our jobs. We are human, after all. But we love to share what our research has revealed to us about the nature of human thinking and behavior. The closest that most academics get to discussions of our work with nonexperts is in the classes we teach.

But psychology is a discipline that cries out for scientists to engage with people beyond the walls of academia. Just about everyone we know has a mind, and almost nobody knows very much about how it actually works. And that's really too bad. We would never let someone build a bridge without learning some physics, or practice medicine without learning biology, would we? So wouldn't it be to everyone's advantage to know a bit more about how brains actually function before we develop an opinion, make a tough decision, or design a curriculum? Shouldn't we first explore what leads us to think and feel and behave the way we do?

And so we opened ourselves up to opportunities to talk to regular people about the human mind. Art began blogging for *Psychology Today*. Bob started working with educators to teach them about how students learn effectively. This outreach has kept us busy: At least once a week we end up speaking to an audience about the wonders of how our minds work.

There wasn't a goal in mind when all of this started—it just seemed like the right thing to do. All that work communicating with other audiences came in handy when Art found himself needing to sell something. It was the winter of 2012, and Art had helped start a new master's program at The University of Texas called the Human Dimensions of Organizations. The aim of the program was to teach folks in business about how people operate, combining studies in the humanities and the social and behavioral sciences. Because the program was new, Art had to look for ways to introduce it to more people. As part of this effort, he reached out to the local public radio station, KUT, because the station manages a music venue on The University of Texas campus called the Cactus Café. Every couple of weeks, KUT hosts a conversation at the Cactus called *Views and Brews*. Art asked if some people from his program could participate, and—much to his surprise—the station agreed.

Because the Cactus is normally a music venue (people like Lucinda Williams, Lyle Lovett, and Robert Earl Keen got their start there), Art thought it would be fun to bring a colleague from the music school along. So he called Bob and asked him to join in.

And that is how the two of us ended up on the stage of the Cactus Café, along with the producer and moderator of the show, Rebecca McInroy, talking about smart thinking and creative problem solving. We had a great time and laughed a lot. And that was that . . . at least for a while.

About a year later, Rebecca was putting together the next year's schedule for the Cactus and asked us to do another *Views and Brews*. Because we like being onstage, we agreed, and when we showed up she asked if we would consider turning our conversation into a radio show and podcast about the mind. Neither of us had ever considered doing a radio show—the expected response would have been a smile and a polite excuse about being too busy. But here is where the personality characteristic "openness to experience" reared its head.

PERSONALITY PSYCHOLOGISTS HAVE IDENTIFIED FIVE DIMENSIONS OF PERSONALity that aptly describe the ways that people differ from one another in terms of their behavior. Not surprisingly, they call these dimensions the Big Five. Each dimension is a continuum, with people on either end being very much unalike with respect to that dimension.

One of the Big Five is *openness to experience*, which refers to people's willingness to try new things. Those who are relatively open will *consider* all sorts of new opportunities. They may not actually try all of them, but they at least are willing to think about them. People who are closed to experience, on the other hand, will generally dismiss new ideas simply because they are new.

People who are closed don't really acknowledge that they dismiss things just because they are new. Instead, they find all kinds of reasons why the new thing is a bad idea: It won't work; it will be too time-consuming; you might make a fool of yourself; it might not succeed, and thus be a waste of time; there are other people who will do it better. You get the idea. New things are a bit scary, so people who are closed respond to that fear by avoiding change, living life as they were living it before.

As luck would have it, both of us are pretty open to experience, so when Rebecca asked us if we'd be willing to do a show, we both smiled and said, "Sure." (As it turns out, we are both pretty extraverted as well, so being in front of a large radio audience also seemed liked fun for us. Extraversion is another one of the Big Five personality traits, and it refers to the degree to which a person likes to be the center of attention in social situations.)

So, a few weeks later, we found ourselves sitting around a table in the beautiful new KUT studios on the University of Texas campus. Neither of us had any clue what we were doing. Our recording engineer, David Alvarez, patiently explained (several times) where we were supposed to put the mike to avoid popping our *p*'s and that we shouldn't tap on the table or kick our chairs. And we promptly forgot much of what we were told. But we did spend a lot of time in the studio talking about topics in psychology—happiness, fear, personality, habits, multitasking, brain games—and we had a blast.

And Rebecca skillfully, brilliantly turned our rambling conversation into seven minutes of coherent fun. The show, which we called *Two Guys on Your Head*, launched in August 2013.

Each week we talk about an aspect of psychology that you probably have thought about before, but we enhance the conversation with actual studies that can help you understand yourself and the people around you just a little better. To give you a sense of what the show sounds like, Bob is the guy who sounds like he is from New Jersey, and Art is the guy who sounds like Bob (probably because he's also from New Jersey). Also, we laugh a lot. In fact, one of the reasons Art likes Bob so much is that Bob laughs at his jokes.

The funny thing about this whole story is that it sounds a lot clearer and more consequential in hindsight than it did when all of this was going on. Life doesn't have a clear narrative as you're living it. You often don't know what the important events are going to be until long after they happen. Calling up a radio station to try to promote a master's program does not seem at first like it will lead to a show on the radio about the mind. But that's how it turned out.

The reason it's helpful to be open to experience is that you never know where things are going to lead. If you try something new, it might just work out, and it might even end up being something you look back on and decide was one of life's turning points. Even if it doesn't turn into anything at all, you might enjoy or learn something from the experience.

Of course, if you are *too* open to experience, you can cross the line from being productively and happily interested in new things to being compulsively novelty seeking. It is a good thing to try on a new idea for size, but you don't have to spend a week at a sweat lodge in Arizona just because your cousin's best friend swears it's a great way to lose weight. The trick is just to truly think through the pros and cons of an action before you decide against it.

IF YOU FIND YOURSELF BEING SOMEWHAT CLOSED TO EXPERIENCE AND WOULD like to become more open, there are several things you can do. The first is to take a lesson from research on regret.

When psychologists started studying regret, they began, as is often the case, by looking at college sophomores (because sophomores are the fruit flies of psychology research—cheap, plentiful, and easy to access). If you ask college students what they regret, they talk about largely stupid things they have done like getting drunk, failing a test, or crashing a car.

Tom Gilovich from Cornell had the great idea to ask people in retirement homes what they regret. When he did that, he found that much of what older people regret is not the things they did, but rather the things they *didn't* do—never learning to salsa dance, never traveling the world, or never learning to play a musical instrument, for example. When people near the end of their lives, they begin to realize that there are things they never did that they are also never going to do.

Once you let it sink in that someday you will regret your in-actions, it's easy to use your remarkable capacity for mental time travel to help you think about what you may regret *not* having done. When faced with an interesting new prospect, imagine your future self in retirement. Ask yourself whether, at the end of your life, you may regret passing up this opportunity. If so, then open yourself up to it.

A second thing you can do is to recognize that your brain has two distinct motivational modes: a *thinking mode* and a *doing mode*. When you are in the thinking mode, you contemplate the upside and downside of a particular course of action. You identify the obstacles that stand in the way of success. You make plans for the future. You think about past successes and failures. When you are in the thinking mode, you do not have energy driving you to act on the world.

When you are in the doing mode, you want to act. You want to engage with the world. You itch to get things done. You get impatient with people around you who are in the thinking mode because you just want to get moving.

Often, when you close yourself off to new possibilities, you are engaging your *doing* mode rather than your *thinking* mode. This may seem like a bit of a contradiction, since in this case the doing mode is leading you *not* to pursue something new. How does that work? Well, your doing mode generates a felt need to act, but because it's generally easier to get moving along paths that are familiar than along paths that are less so, the action you choose to do is something that is *not* new.

So when you're tempted to close off a new possibility, consider giving yourself permission to think over the new idea—that is, wait to decide. Live with it for a few days. Even if the idea seems strange and uncomfortable at first, it may grow on you over time. Rather than dismissing it, just let it hang around and see if it becomes more interesting or exciting.

When the fear of the unfamiliar dissuades you from striking out on a new path that in some ways may seem attractive, it is worth asking this question: *What is the worst thing that could happen?* Many of our fears are more intense than the dangers we perceive actually justify. Consider public speaking. Talking in front of an audience creates fear in so many people that psychologists routinely use the threat of giving a speech in front of other people as a way to induce stress in experiments. The two of us speak in front of others for a living, so for us that fear is only a distant memory. In fact, Art has admitted that he finds it harder to sit in the audience than to give a talk (there's that extraversion). How do you accomplish that switch?

The simple answer is practice. But what's the active ingredient in this practice? In part, practice helps you to improve your ability to give talks, which gives you more confidence. Even if you are more introverted than either of us, you'll benefit a lot from practice in giving talks. Practice helps you to realize that there isn't really that much to be afraid of, in that you go through the experience of speaking without many negative consequences. After all, unless you are a politician at a press conference, there is very little you can say when speaking in front of others that will have lasting repercussions. The actual danger in giving talks is quite small relative to your beliefs about the danger. If you make a mistake, people may laugh for a moment, but then that moment is gone. In fact, most listeners are actually quite forgiving of the mistakes speakers make.

Of course, perceived dangers aren't *always* disproportionate to actual dangers. If you are given the chance to go bungee jumping, you may very well chicken out. Art won't even climb on a roof, let alone jump off a high structure with elastic tied to his feet. It's not that he's afraid of heights; he's just afraid of pain and death. That seems reasonable, especially since the thrill of diving headlong toward the ground doesn't seem to compensate for the potential crash. The downside of bungee jumping really is significant, however unlikely, so Art feels quite justified in his fear. (Which is not to say that *you* should avoid bungee jumping—only that Art won't be going with you.)

In general, though, the contemporary, industrialized world is pretty safe (even bungee jumping, apparently). So before you close yourself off to a new opportunity, ask yourself whether the only thing you have to fear is fear itself. (FDR knew a good line when he heard one—if we keep using it, maybe it will catch on.)

We can feel pretty confident that, in the end, we're likely to enrich our lives by being open to new opportunities. The people you think of as successful and productive are generally people who were open to new experiences rather than tied to beliefs about how their lives were supposed to proceed. And if you find that you're not the kind of person who is open by nature, and you would like to become more so, there is also a lot you can do to push yourself to *act* more open than you feel. Which in turn will lead to your actually *feeling* more open than you once were.

IN ONE OF THE EARLY EPISODES OF THE SHOW, BOB GAVE A PITHY SUMMARY OF what we had been talking about and then said, "You know, we ought to cross-stitch that on a pillow." Ever since then we've been using that line to describe memorable insights we gathered from our little odysseys into human psychology. Throughout the book, we'll be tossing out ideas for your next cross-stitching party, or custom mug motif, or tattoo, or T-shirt silkscreen design, or whatever floats your boat.

So for this chapter, here is a nice little aphorism to summarize the advantage of openness:

Can we really make ourselves happy?

ABOUT TWENTY YEARS AGO, THERE WAS A PUSH IN THE FIELD OF PSYCHOLO-gy to shift focus away from studying the things that go wrong in people's lives, like stress, and toward studying the things that go right, like well-being. This movement—aptly named *positive psychology*—was championed by Martin Seligman, who was then president of the American Psychological Association, and Ed Diener, a professor at the University of Illinois.

It was an especially interesting shift for Seligman to start studying what makes people feel good. One of his most prominent early lines of research explored *learned helplessness*, a behavior that develops after organisms (remember, we're all organisms) are repeatedly exposed to inescapable, painful events. What's learned from this experience is that trying to escape is pointless, and the remarkable result is that we organisms eventually give up and don't attempt to escape *even when it's possible to do so*. We'll have a bit more to say about learned helplessness later in the book.

Seligman, Diener, and other psychologists recognized that in order to help people lead lives filled with happiness and a sense of well-being, it was necessary to first understand what happy, satisfying lives look like. Rather than focusing on what makes people miserable, they asked, *What is it that makes people happy?*

Research findings about happiness are interesting and in some ways counterintuitive. One of the most important findings is that individual happiness remains fairly stable over time. Some people are pretty happy (or satisfied with their lives) most of the time, while others are not so happy most of the time. Even with the vicissitudes of life's events and the ups and downs that we all experience, everyone seems to have a level of happiness that remains relatively constant over the course of their lives. Researchers call these general levels of overall happiness *set points*.

Even though various life events, like getting a pay raise or ending a long-term relationship, can make us feel more or less happy in the short term, all of us tend to return to our set points over the long term. A death in the family understandably brings sadness that may last for weeks and months (or longer). Winning the lottery brings exhilaration and joy in the weeks and months following the windfall. But in most cases the subsequent events that we experience, combined with our own set points, tend to bring us back to the happiness level that we most often experience.

We could be forgiven for believing that our happiness is wholly dependent on what happens to us—what we gain and what we lose, what we accomplish and where we fail—and that the combination of life's events is the sole determinant of our sense of well-being. But this is not actually how happiness is determined.

A study by the psychologist Dan Gilbert and his colleagues illustrates this point. They looked at college assistant professors and assessed how decisions about their being promoted in rank and earning tenure affected their happiness.

Tenure brings with it a tremendous amount of job security, and earning a promotion and tenure is a very big deal, with lasting life consequences. You might think that if somebody told you that you could keep the most awesome job in the world for as long as you wanted, you would be happy for the rest of your life. And conversely, if you were told that you couldn't have that job anymore, you would feel devastated for a very long time.

One group of professors who were about to be evaluated for tenure were asked to predict how happy they would be in the months and years after their tenure decision. They made predictions about how they would feel if they were awarded tenure and how they would feel if they were denied tenure. It was no surprise that the professors predicted they would be happier if they got tenure than they would be if they didn't, and they estimated that the effect of being denied tenure would last roughly five years.

A second group of professors who had already been evaluated for tenure—some who earned tenure and others who did not— were asked about their actual happiness after their cases had been evaluated and the decisions made. Did earning tenure make people happier? Not really. Those who had been denied tenure were just as happy, on average, as those had been awarded tenure. Sure, they were upset for a few months, but this major life event had much less influence on their lives than they had expected it would. Observations like this are probably the root of pillow-worthy adages like *This too shall pass*.

This does not mean that our overall levels of life satisfaction can never change, but studies of happiness among large groups of people over long periods of time reveal two interesting results that we mentioned earlier. First, from one year to the next there are fluctuations in happiness that are the result of changing circumstances. Second, as time passes after the emotional events, whether positive (like a great accomplishment) or negative (like a serious illness), we all tend to return to our set points, although it may take time for that to happen.

Findings about changes in long-term happiness may seem a tad depressing, especially for people who tend not to be very happy. When there are long-term changes in overall life satisfaction, those changes tend to be negative rather than positive—that is, we are more likely, on average, to become less happy over time than we are to become happier. This is because we are more likely to encounter life circumstances that cause problems and reduce happiness than we are to encounter favorable circumstances that improve our lot. Although the tendency for all of us is to return to our relative set points following positive or negative events, prolonged illness, unemployment, or death of a spouse, for example, can lead to long-term decreases in how happy we feel.

One factor that *does* tend to increase life satisfaction over the long term, however, is marriage. Even though movies and television often portray marriage as a source of stress, marriage actually tends to be a source of stability. That stability also helps us develop healthy long-term habits, which in turn help prevent illnesses that can decrease life satisfaction.

So what can we do to make ourselves happier? Perhaps the most important thing is to remember that happiness is not dependent on your achieving some momentary life goal. We see lots of people who have put their happiness on hold while they slog through days that are generally unsatisfying, all the while expecting that when they eventually reach their goal (graduating, getting married, finishing a book, having a child, buying a house), their happiness will be switched on for good.

Art tells a great story about hiking the Grand Canyon as a teenager. After reaching the bottom of the canyon, he began the long trek back to the top, a path that involves lots of long switchbacks that slowly lead back to the rim. You walk up the trail for a while, and then it turns back in the opposite direction and you walk some more. You can't see much beyond the switchback, so it is hard to figure out how much progress you've made. You feel that the top must be right around the next bend. That expectation continues for a long time, until you finally (and happily) reach the top.

Expectations about life's happiness may feel just like that. You figure that happiness is just around the next bend. It will happen when you get accepted to an elite college. Or when you graduate and get a job. Or maybe when you get that promotion. Or perhaps when you finally find a partner. Because you're basing happiness on some future event, you keep delaying when you will really have a chance to be happy until you reach what's around the next bend.

Too bad, because the thing is, you are living right now. And you're in the Grand Canyon! All of that uphill hiking may be a challenge you suffer through until you reach the top. Or you might think about and take advantage of all the beautiful things you experience as you pant and sweat your way up the trail.

Rather than postponing joy until you reach some goal, only to recognize that the joy of reaching it is rather short-lived, you can focus on what there is to love about what you're doing right now, today. Yes, there are things that we have to do and sacrifices that we need to make for the sake of some future goal. But if your life is little more than an unending string of tasks you are not enjoying for the sake of some hoped-for future, it is time to reevaluate the things you are doing each day and find something you can do today that will give you some enjoyment and fulfillment. There is an old bumper sticker that says, *He who dies with the most toys wins* (note the gender designation), but while the ones with the biggest piles of toys may win the race for accumulating the most toys, they won't necessarily be happy. Plus, they'll be dead.

It's not surprising that people who enjoy the work they do and think of it as a calling are happier than those who experience lots of stress and anxiety at work without any greater sense of purpose. People who feel trapped by life circumstances, like a crummy job, understandably feel unhappy.

Bob sometimes meets with undergraduate advisees who have come to view school as little more than an endless source of stress and frustration, and he suggests that they drop out for a while—advice that's met with a bit of shock, coming from a faculty advisor. Remember that these are privileged people who have lots of choices in life. Most of the students who hear this advice don't actually leave school. But recognizing that they *could* drop out and that attending college is their *choice* helps them to rethink why they are there. They feel less trapped by the circumstance and more in control. In general, it's valuable to remember that we often have more options than we recognize.

WE'LL END THIS CHAPTER ABOUT HAPPINESS EMPHASIZING THE IMPORTANCE of connecting with other people. Studies of happiness show that loneliness is a big negative predictor of happiness (which is a roundabout way of saying that lonely people are generally unhappy people). If you don't feel connected to the people around you, you feel worse than when you feel a strong connection. This is true irrespective of whether you're in a close romantic relationship.

Part of the problem, of course, is that unhappy people often don't want to reach out to the people around them. So the correlation between loneliness and being unhappy is exacerbated by the fact that feeling unhappy leads people to be less sociable. But the number of people you spend time with is something you can control. If you feel lonely, you can call a friend or perhaps a family member. There are also lots of great ways to meet new people with similar interests without having to spend a dime. Community organizations, religious institutions, and various social causes offer lots of volunteer opportunities that connect us with other committed individuals.

It turns out that even random conversations can make you a bit happier. Nick Epley and Juliana Schroeder have done some cool studies focusing on commuters. Most commuters on public transportation think that they will be happiest on their commute if they sit alone and work or read, and many live in mortal fear that a boring person will sit next to them and engage them in a long and pointless conversation.

Epley and Schroeder asked some people to engage in conversations with strangers as they commuted to work. Others were asked not to talk to other people. It turns out that commuters enjoyed the conversations and enjoyed their commutes more than did those who were asked to keep to themselves. It may be that many of us are missing lots of little opportunities to make each day a little more pleasant.

You see what we're getting at here. A big chunk of our long-term happiness reflects our genetic predispositions, and there's nothing to be done about that. But there are many things within our control that can lead to a more satisfying and happier life. This is not to ignore the fact that all of us face varying levels of challenges. It is easier for some of us to be happy than it is for others. But happiness is not entirely beyond our control, and by taking the opportunity to appreciate our surroundings and getting to know one another, we can potentially improve our overall life satisfaction. In sum:

Sometimes happiness just comes to you;

SOMETIMES YOU HAVE TO GO AND GET IT.

How do
we catch
a liar?

PEOPLE LIE ALL THE TIME. EVEN THOUGH SOCIETY TALKS ABOUT THE value of telling the truth, there are many situations in which it is probably best for your social life to lie. You might compliment a person's outfit, even though you don't like it much. You might invent a fictitious obligation with family to avoid going to a party that you think will be boring. These little white lies help to maintain pleasant social interactions with other people. Of course, people tell bigger lies, too. People lie to cover up situations in which they have broken the rules, or to blame others for their mistakes.

Although what we refer to as little white lies may be of little lasting consequence, when the consequences are high, it is valuable to be able to figure out whether someone is lying to you. Over the years, lie detection has become a big business. And the ugly truth is that it can be quite difficult to distinguish between people who are lying and those who are not.

There are several different approaches to lie detection. Some of them assume that telling a lie creates stress or mental energy (often called *arousal*), because there is tension between what people know to be true and the information they present that is false. The liar may also fear being caught, which adds to the arousal. There are several physiological signs of stress and arousal that are hard (or impossible) to control. Heart rate speeds up under stress, and people take shorter breaths. In addition, there are changes in skin conductance that reflect small increases in the amount of sweat produced.

These measurements are the basis of polygraph tests. The basic idea is to take advantage of these physiological changes to assess whether particular statements that people are making are lies.

There are several problems with using arousal as a measure of lying, though. For one thing, when people repeat a lie often enough, it becomes psychologically the same as the truth. There is no longer a tension between the true statement and the false one. When that happens, there isn't a physiological reaction to the lie.

A second problem is that some people find the procedure of being connected to a polygraph very stressful. If they are asked about sensitive topics, they may show enhanced arousal and give a response that looks like a lie. That is because the polygraph measures an *expected consequence* of lying—not the lie itself. If the expected consequence doesn't occur when someone lies, the test doesn't work. If the arousal response occurs for some other reason, the test wrongly indicates lying. That is why polygraph tests are no longer admissible as evidence in court.

As a psychologist, Art often gets asked about particular theories related to catching liars. Friends approach him with all kinds of ideas: liars blink more often than people telling the truth; liars look up and to the left; liars laugh while they are talking; liars squeak when they talk; liars won't make eye contact with you while they speak.

Some liars undoubtedly do some of these things, some of the time. As do some people telling the truth. Consider eye contact for a moment. It seems to make sense that people might have trouble making eye contact when they are lying because they may feel embarrassed, as though the person they are lying to would be disappointed in them if he found out the truth. Indeed, some liars probably do have trouble making eye contact.

But in fact, almost everyone has some trouble making eye contact while they are speaking to someone else. The human face is complex and causes a lot of brain activity when you look at it. Your brain wants to interpret the subtle expressions that a person is making to get a sense of how that person is feeling.

But that brain activity can actually interfere with the work of turning your thoughts into sentences you want to speak. As a result, many people look away from the person they are talking to while they are speaking and then make eye contact again when they are listening to the other person speak. Test this out: Next time you're having a conversation, pay attention to when the other person looks directly at you. You'll most likely find that her eyes wander away from your face when she is speaking.

None of the physiological cues that are associated with lying are perfect indicators, but they do convey some information. Most of the time, people are unable to use these cues to catch liars effectively. That said, there is some evidence that people are sensitive to the cues others give off when lying—they just can't use them explicitly very well.

Most supposed cues for lying are actually quite poor indicators of lies—although not all of them. For example, people often display more stress in their voice when they are lying than when they are telling the truth. But it turns out that if people watch someone telling a story, they are often a better judge of whether that person was lying when asked about it the following day than they are immediately after seeing the person. The idea is that the subtle cues that someone is lying are overshadowed by people's explicit beliefs about liars while they are engaged in a conversation. Later, though, those explicit beliefs fade, and the more subtle cues people can pick up on (like vocal stress) play more of a role.

Some aspects of behavior also help to distinguish between lies and truth. One big difference between fact and fiction is in the language people use to tell their stories. In his book *The Secret Life of Pronouns*, James Pennebaker explores studies that analyze the words employed by people telling the truth versus those who are lying about similar topics.

There are several broad differences in the way that liars and truth tellers discuss events. One difference is that liars say less overall than truth tellers. If you are telling the truth, the details of what happened are obvious. If you are lying, it is not easy to conjure up lots of details. Interestingly, truth tellers talk less about their emotions than liars do. That is because when you recall a real memory, you begin to reexperience some of the emotion from that event. As a result, that emotion feels obvious to you (and would be obvious to anyone watching you). If you are lying, though, you don't really experience that emotion, so you describe it instead. Truth tellers also talk about themselves more than liars, because people telling the truth are more focused on their own memories than liars are (who are also thinking about how their story is being perceived by others).

NEVERTHELESS, IT IS HARD TO USE THESE OBSERVATIONS ALONE TO CRAFT A technique for catching liars, because there are lots of differences among people in their overall use of language. For example, we both talk a lot, so we use a lot of words when we are telling the truth. Some other people tend to be succinct in the stories they tell. Similarly, some people talk about themselves a lot, while others don't. In order to use the facets of lying outlined above to catch a liar, you actually need to compare a particular story a person is telling to situations in which he is known to be telling the truth.

Because liars have not experienced the situation they are describing, they are able to say less about it than truth tellers—they don't know as much about the situation. Suppose, then, you asked liars and truth tellers questions they didn't expect to hear. For example, suppose Bob is at a party and he starts regaling people with stories about being on the radio. People could ask him questions about what the studio is like, who engineers the program, or whether he gets to meet some of the great reporters at the station. Bob would have answers to all of these questions based on his experience, even if he didn't expect someone to ask those questions.

But suppose someone claimed that she went to a restaurant she had never been to before. She would have trouble giving answers about where the bar is in the restaurant, what street it is on, whether it has a parking lot, what kind of lighting it has, or whether it is loud. There are lots of incidental pieces of information that people have when they are telling the truth that they simply don't have if they are lying.

In a fascinating study, researchers trained airport security screeners either to use traditional cues to lie detection (like where people are looking) or to use an interviewing technique in which they asked people questions about common knowledge that someone would know if they were telling the truth about their travel. Then, as a way of testing the effectiveness of the screeners' procedures, people were sent through security and asked to lie about their travel intentions. The testing took place over a period of several months. Some people were sent through several times to see if they could improve their ability to fool security screeners. These people were paid for their participation in the study and were given more money if they were able to deceive the screeners than if they were not.

In the end, screeners trained on the traditional cues for lie detection caught the liars about 5 percent of the time—not a great record—while those trained on the interviewing technique succeeded 70 percent of the time. And people who went through security multiple times did not get any better at fooling the interviewers.

The real problem with lie detection is that people have bad theories about what gives liars away. They focus on the nervousness they expect people to feel when they are lying. Instead, focus on memory. People who are lying did not experience the events that they're lying about, so they cannot recall the same kind of information about their experience that truth tellers do.

Should we play brain games?

OKAY, MAYBE WE SHOULD ELABORATE.

Getting smarter is big business these days. We live in what management guru Peter Drucker called the *knowledge economy* (according to Art, it is required to say "management guru" whenever you say Drucker's name). That means that most of the best jobs involve learning and using information and pushing the envelope of our understanding of the world.

So it's no surprise that we crave simple exercises that will make us smarter.

There are a number of so-called brain games currently on the market that promise to do just that, and in addition to their promises, they share a number of design features: They don't take long to play; they are not related to any content you know, so they feel more like games than like schoolwork; they are engaging but often not very enjoyable compared with watching a movie or reading a book.

Brain games are intended to exercise your mind in the same way that physical activity exercises your body when you visit the gym or go for a run. Lifting weights, jogging, or working out on an elliptical trainer help us stay in shape without spending an inordinate amount of time. They strengthen parts of our bodies in a specific context, outside of normal daily activity. And for many of us, the activities are not really fun in and of themselves, which is why many runners wear earbuds and why elliptical trainers and treadmills come with video screens and headphone jacks.

Yet the physical activity we do in the gym actually accomplishes what we expect it to. Physical exercise *does*, in fact, strengthen your body, develop your muscles and your cardiovascular system, and increase physical endurance.

But what about brain games? Can they really make us smarter?

Sadly, brain games have several fundamental limitations that make this unlikely. There are many studies demonstrating that performance on certain kinds of psychological tests is related to good performance in school. Tests of *fluid intelligence* examine our ability to do logical problem-solving tasks, and performance on these tests is related to how well we do with academic challenges that involve problem solving.

Performance on fluid intelligence tests is related to a mental capacity called *working memory*, which reflects the amount of information we can hold in our conscious minds at one time. Working memory is where we compare, combine, and evaluate ideas. The larger our working memory capacity, the better we're able to process information. Some people can easily recall a list of seven digits after hearing the list just once; others can easily remember ten digits. Those who can remember ten digits do better on tests of fluid intelligence than those who can remember only seven.

The logic of many brain games is that if you exercise and stretch your working memory capacity, doing so will increase your fluid intelligence, which in turn will improve your performance in problem-solving situations like those you encounter in school or at work. There are other types of games that are designed to strengthen other capacities, like spatial reasoning or executive attention.

Sounds reasonable enough, but it's not quite that simple. Mental capacities like working memory have to function in a coordinated fashion with lots of other systems in the brain that retrieve information from memory and control your attention. Brain games fail in part because they treat thinking as if it involves a bunch of separate mechanisms that can be improved independently. Just doing a test that aims to improve one capacity does not guarantee that this will improve the *coordination among capacities* that makes you smarter more generally. It's like exercising only your calf muscles and expecting to become a faster runner.

A second big problem with brain games is that they are focused on the wrong aspect of thinking. Back at the dawn of cognitive psychology (way back in the late 1950s), two of the titans in the field, Allen Newell and Herb Simon, explored methods for solving problems. They pointed out that people have lots of general-purpose strategies for solving problems that can be used in many situations. For example, the strategy of *working backward* says that you start with the goal you have in mind and then try to figure out a series of steps that gets you from that goal to the situation you are in right now. This kind of strategy is now often referred to as reverse engineering. The strategy of *hill climbing* involves starting with the current situation and taking steps to reduce the difference between that situation and your goal.

Newell and Simon called strategies like these *weak methods* of problem solving. They don't require much knowledge to be usable, and even though they can be applied in lots of situations, they often don't work very well. They're go-to strategies when all else fails.

Strong methods of problem solving, in contrast, are based on domain knowledge—that is, knowing stuff about the problem you're confronting. Art has no earthly idea how cars work, so if he gets in the Art-mobile and it makes a funny noise, he has to bring it to the mechanic. The easiest problems for the mechanic are like the ones he (Art's mechanic happens to be a "he") has seen before. But even when the mechanic encounters an unfamiliar problem, he knows a lot about cars, so he can quickly do some diagnostic tests based on the way cars work, find the problem, and fix it. He doesn't have to rely on general strategies because he's an expert on cars.

If Art brought his car to Bob (who knows about as much about cars as Art does), the two of us would be stuck. We don't have any strong methods of problem solving we can use in this situation because we have no knowledge about how cars work. Worse yet, we don't know enough about cars to even effectively apply many of the weak methods of problem solving, like working backward or hill climbing. We could try very general strategies like wiggling cables or looking for loose screws, but all of this (very weak) problem solving would probably just make things worse. Even when weak methods work, they take much longer than strong methods.

We know that intelligence isn't just about basic mental capacities. The type of intelligence that involves knowing something about the domains you're working in is called *crystallized intelligence*. This term has never made sense to us, because the word *crystallized* makes it seem as though knowledge makes you less flexible (more crystallized), but actually the opposite is true. Knowledge about what you're working on makes you *more* flexible in your ability to act.

NOW BACK TO BRAIN GAMES.

Suppose you wanted to create a brain game that could be played by millions of people. It's impossible to estimate what knowledge these millions may have, so it's better to build a game that doesn't require players to have any particular knowledge, like: There are two pints in a quart; force equals mass times acceleration; Texas is the Lone Star State. That way, anyone who picks up the game can use it.

Creators of intelligence tests confront a similar challenge. They can't know for certain what future test takers will know (or need to know), so they build tests around tasks that don't require specific knowledge. Builders of brain games and intelligence tests go out of their way to avoid any kind of content knowledge requirement and instead focus on factors that relate to weak methods of problem solving.

(As an aside, even tests that don't require any specific knowledge do require some understanding of what tests are all about. But that is a topic for another time.)

If you really want to improve your ability to think effectively, optimizing your skill with weak methods of problem solving is not really going to get you there. Instead, you need to develop expertise that will support strong methods of problem solving—that is, you need to learn stuff (and thus improve your crystallized intelligence).

It's not that the fundamental intent of brain games is unreasonable. You *can* make yourself smarter—but not by playing brain games. What brain games help you do mostly is improve your performance on brain games. If you enjoy the activity, fine. But if your goal is to become smarter, you'd be much better served reading a book (hey, you're already doing that right now!), learning from YouTube® videos or online courses, or subscribing to magazines with long and detailed articles on varied topics, so that you learn enough to really understand something new. The knowledge you pick up (your expanded crystallized intelligence) will contribute much more to your ability to think and work smarter in life than will any small improvement in working memory you may get from playing a game.

We'll close with a interesting, non-intuitive thought: If you have a spare twenty minutes in a day, the most important thing you can do for your brain is to take a nap. The modern world has created a huge number of sleep-deprived people. The benefits of a twenty-minute nap far outweigh whatever you will get out of a game you play for the same amount of time. And it's a nap!

And that leads to this chapter's pillow-worthy adage:

5

Do stories help us remember?

OOOH. I'VE ALMOST GOT IT. WAIT. WAIT. GIMME A SEC. WELL, maybe not . . .

Trying to recall facts often leads to experiences just like this. We know we know it, we know it's in there somewhere, we've almost got it, but we can't quite seem to find it in the vast expanse of our memories.

It may be that in the age of Google™ and IMDb®, we have fewer such experiences because our initially unsuccessful memory searches often end very quickly when we whip out our phones. Why continue to struggle when we can look it up right now?

But for those of us who enjoy trivia games, one of the best and most fun things about them is that they trigger lots of so-called tip-of-the-tongue experiences. In a typical tip-of-the-tongue state, we not only feel that we know the information, we can even recall some aspects of what we're trying to retrieve. We might be pretty confident about the first letter of a word or whether the word is long or short, for example, even though we can't quite recall the word itself.

Here's how this works. When you receive a cue that reaches

into long-term memory (like a trivia question), individual memories that relate to the cue leap up (are *activated*) and compete to be selected. This competition also leads some memories to damp down (*inhibit*) others in the mix. Because trivia questions are about trivia (that is, information of little importance or value that is rarely accessed), the relationship between the question and the answer is usually relatively weak, so any competitor in memory provides enough inhibition that the memory you are trying to recall can't quite surface. You may get its outline, but you may not be able to quite retrieve all of it.

In many cases, when you fail to recall a word or name, it seems that the harder you try to remember, the more difficult it becomes to locate what you're searching for. It's like repeatedly running headlong into the same dead end. Ouch.

The way to get past an insistent state of almost-having-it is somewhat counterintuitive: Stop trying. Allowing your brain to stick in the same cul-de-sac with all those competing memories seldom leads to a successful outcome. But when you direct your attention elsewhere—think about something unrelated, take a walk, do a small chore—your mind settles back down. When you return to the question later, the right answer has another chance to come to light without being beaten back by competitors.

Often, the right answer just leaps out unbidden, even while you're thinking about something else. Showers and baths are places where you often get these reminders, because while you're bathing there is very little going on, very little that you have to really think about, and very few other cues to retrieve things from memory. (Unfortunately, these aren't places where it's easy to write down what you recall.)

Art distinctly remembers playing a trivia game in college and being asked the name of the treaty that ended World War I. He had done fairly well in history class, but the name of the treaty completely eluded him during the game. When he was told it was the Treaty of Versailles, Art knew that he had been exposed to that information in the past. Indeed, he was pretty sure he had successfully written that in response to exam questions.

So what makes some things easy to recall and others difficult or impossible?

The more that a given piece of information is connected to other pieces, the easier it is to remember. You are unlikely to forget the name of the US president during the Civil War, for example, because so many different things are associated with Abraham Lincoln. All of these connections make it easy to remember him because they provide multiple pathways to the name. The Treaty of Versailles is harder to remember (at least for Art), because it is not connected to as many other pieces of his knowledge.

And this leads us to why stories are so much more memorable than individual facts: Stories provide lots of interconnections among individual pieces of information. These interconnections make stories and the facts embedded in them easy to recall, which is why even very young children can recount long and elaborate stories without breaking a sweat. The facts that make up a story are connected to other related facts, which are connected to still other facts, and each part of the story triggers the recall of other parts.

WE BOTH LIKE TELLING JOKES, AND EVEN JOKES WE HAVEN'T TOLD IN A LONG time seem to roll off the tongue with very little effort. One big reason we're able to remember lots of jokes is that each one involves a few characters and situations that are profusely interconnected in our memories. The content of every joke is a setup for the punch line at the end. We remember all of the elements that make the punch line funny, and that helps us recall the structure of the entire joke.

These interconnections are not the only thing that helps make stories memorable, though. You have heard many stories in your life, and as a result you have developed *schemas*, which are structures or outlines of what good stories entail. Because of these schemas, you have developed expectations about what's likely to happen in stories, and you tend to remember things that relate to your schemas.

In a classic study from the early 1970s, John Bransford and Marcia Johnson had participants in an experiment read a story that was preceded by a title suggesting the story's schema. For example, one story was titled "Watching a Peace March from the 40th Floor." (We did say this was the early 1970s.) Most of the story was about crowds moving about, television cameras, and speeches. In the middle of the story, however, was the strange sentence: "The landing was gentle, and the atmosphere was such that no special suits had to be worn." Most people who read the sentence didn't really understand it because they couldn't make it fit with the rest of their schema for peace marches. As a result, when they were later asked to recall the story they had read, they didn't remember that particular sentence at all.

Another group of participants read the same story but were told its title was "A Space Trip to an Inhabited Planet." This group was perfectly able to understand the strange sentence when they read it, and they also were more likely than the participants who thought they were reading about a peace march to recall it.

When you read or hear stories and when you experience events in your life, you understand them in light of the schemas you have developed over time. Not only do those schemas affect which parts of an event make sense to you, but they also influence what you remember about it later.

When you experience unfamiliar situations, you tend to pay more attention to aspects that fit with your existing schemas than to aspects that don't. Likewise, you tend to remember the details that are consistent with your schemas. Things that don't fit often drop out of memory. Events that happened in an unusual sequence (compared with the sequence in the relevant schema) can get switched in your memory back to an order that you've more typically experienced. When you recall past events, you may even insert details into your recollections where they don't belong.

WITH THE PASSAGE OF TIME, ALL OF YOUR MEMORIES TEND TO MORPH IN WAYS that resemble typical schemas. Why not simply record events as they happen and store them with high fidelity?

The first step in understanding the answer to this question is to recognize that the purpose of memory is to help us interpret the world *in the future*. Everything we encounter in the present engages schemas that help us predict what will happen next.

Most of the time, when you need to recall past experiences or stories you have been told, it isn't necessary that you get all of the details right, as long as you can remember enough to do a reasonably good job of predicting what will happen in new situations. So unlike a video recording, your memory can make systematic mistakes about what happened during past events—mistakes that most often have no negative consequences.

When you recall a memory, it feels as if you're simply accessing complete buckets of information that you can then think about. Consider your memory of having breakfast with family on a Sunday morning. The experience of having breakfast includes lots and lots of pieces: tastes, images, smells, sounds, conversations, movements, and emotions. And when you think of having breakfast, all of those pieces seem to come neatly knitted together. What you don't realize is that at the time of recall, all of those pieces have to be collected from the various locations in the brain where they were stored and then *put back together* to form a memory. We do this each time we recall anything.

Schemas form a template that helps our brains glue pieces of memories together. Because your memory doesn't faithfully record everything that happens (not by a long shot), there are a lot of details you need to fill in to make the memory tell a coherent story. You use one of your schemas to fill in those details. And as a result, you will recall things that didn't actually happen.

You also tend to combine information from different events that are related to a common schema when doing so helps you make predictions for the future. From an evolutionary standpoint, this makes sense. The creatures who were best able to make predictions are the ones who survived. It didn't actually matter if those predictions were based on a single instance or multiple experiences that were woven together. However, this tendency can also lead to serious mistakes that are difficult for you to recognize, particularly in eyewitness testimony situations in which we rely on the accuracy of people's memories.

Elizabeth Loftus and her colleagues showed people a film of a collision. After viewing the film, the research participants wrote about the accident and answered questions about it. A question about the speed of the cars was worded differently for different groups of participants. Some were asked about how fast the cars were going when they *hit* each other. Others were asked how fast the cars were going when they *bumped* (which suggests a slow speed) or *smashed* (which suggests a high speed) into each other.

Participants who heard a word suggesting high speed (like *smashed*) indicated that the cars were going faster than those who were given a word suggesting low speed (like *bumped*).

Participants were then asked whether there was broken glass after the accident. There was no broken glass shown in the film, yet the participants whose questions included the high-speed words and who estimated higher speeds of impact "remembered" seeing broken glass at the scene. Participants whose question included the low-speed words and who estimated slower speeds, on the other hand, did not remember seeing broken glass.

Viewers of the film tended to recall what seemed reasonable given the schemas that were activated in their memories. Smashing cars usually produce broken glass. Bumping cars, not so much. This study and many others like it illustrate that you often incorporate information from different modes of experience (like vision and language) and combine them into a single story when you are recalling information.

All of this is to say that when you understand a story, you use your knowledge of the stories you've encountered in the past to help you. This knowledge—in the form of schemas—also affects what you remember later on. When you retrieve a story from memory, your recollection may include inaccuracies, both because you made predictions using schemas as you were first experiencing the story and because those same schemas affect what you include in your reconstructions of the memory.

Even with those inaccuracies, many of which are trivial, the story is useful to you in knitting together disparate components and forming something that seems to make sense. The connectedness and sensibleness allow you to recall more than you could if you tried to stuff disconnected bits of information into your head.

Understanding that recalling memories is, in fact, an act of reconstructing stored components of past experiences suggests that you should be a little skeptical about the fidelity of your memories, even those (perhaps especially those) about which you feel the most certain.

But remember:

MEMORIES
don't have
to be
ACCURATE
to be
USEFUL.

Is pain open
to interpretation?

I F YOU LISTEN TO A LOT OF MUSIC, AS THE TWO OF US DO, YOU'VE PROBABLY noticed that there are lots and lots of songs about pain. Some of that pain is physical (as in songs about hangovers), but most modern-day troubadours sing about emotional pain, often caused by love and loss. Pain is a universal feature of the human experience, but what is pain?

The sensation of pain from physical injury starts in pain receptors located throughout your body that are activated when you are cut, burned, stretched, bent, or broken in ways that are dangerous. When pain signals reach the spinal cord, they initiate other signals that zip back downstream to the muscles, causing them to contract and disengage from whatever is causing the pain. Remarkably, this reaction happens even before the signals that continue upstream to the brain have had a chance to create a conscious experience of pain: You have pulled your hand away from the hot stove before you consciously feel the burn. In this way, the body acts to protect itself as quickly as possible.

Once pain messages reach the brain, the pain is mapped to a part of the body. Most of the time the mapping is quite accurate: Prick a finger on a needle, and you perceive pain where the needle pricked. At times, though, the map is not so accurate, especially when the source of the pain is in a part of your body that doesn't typically experience pain. In these instances pain is *referred* to another part of the body. This is why heart-attack victims in movies often clutch their left arms even though the heart resides in the center of the chest.

Because pain is processed in the brain, it can even be experienced in limbs that are no longer attached to the body. It is common for people who have lost limbs as a result of accident, war, or illness to continue to perceive feelings in their lost limb long after it is gone. One of the worst feelings is called *phantom pain*, which is a real sensation of pain in a missing limb. Some sufferers report that they feel as though they are having a spasm in the arm that is causing them to curl their missing hand into a fist, even describing the pain of their fingernails digging into their palms. How can this happen?

Art's favorite among our many human senses is *proprioception*, the sense that monitors the location of your limbs in space. You don't have to look at your arms and legs to know where they are. You can *feel* where they are. For an amputee to experience that his missing hand is curled into a fist, proprioception has to tell him where that limb "is." Phantom pain is especially awful and difficult to stop, because you can't uncurl a fist if it no longer exists.

Neuroscientist V. S. Ramachandran has done a lot of work with amputees, and he developed an ingenious way to help people who experience phantom pain. He arranges a series of mirrors so that when amputees look at where a missing limb *would* be, they see a reflection of their attached and healthy opposite limb. While looking where the missing arm would be (and seeing a reflection of the intact arm), the amputee is told to imagine opening and closing the missing hand while at the same time opening and closing the healthy hand, creating a visual sense that the missing hand is opening and closing paired with the physical sensation of doing so. This process actually helps alleviate the phantom pain.

How does that work?

The brain has two detailed maps of the body: a sensory map and a motor map. The sensory map helps the brain to figure out where in the body you are feeling touch, heat, or pain. The motor map helps the brain figure out which body parts to move. When someone loses a limb, the sensory and motor maps for that limb remain in the brain. Anytime the sensory area of the missing limb is activated, it is perceived as a sensation coming from the missing body part that was mapped to that location in the brain. Ramachandran's technique works because it gives amputees visual information about their missing limb. What they see creates activity in the sensory and motor maps for the missing limb, which can relieve the pain. But the brain is a very expensive organ to run, so the cortical real estate taken up by maps that are not serving any purpose tends to diminish in size over time. Eventually, the maps for missing limbs become smaller, as those regions of the brain start to take input from other parts of the body.

IT IS INTERESTING TO CONSIDER HOW MUCH CONTEXT AFFECTS THE INTENSITY of pain. A barefoot ten-year-old who whacks his toe on a chair at home may scream in agony until his mother comes to console him (ice cream often functions as an effective painkiller). But when that same ten-year-old is thrown to the ground by three friends while playing football, he's likely to pick himself up and get ready for the next play, scraped knees notwithstanding.

Experimental evidence shows that what we think of as pain

has both a physical component—the sensation itself—and an *affective* component, which reflects how much agony the pain is causing. Morphine and other opiates are interesting drugs, because they don't actually dull the sensation of pain; they just make the pain more bearable. In a study of people who suffer from chronic pain, participants were given IV drips that dispensed either morphine or a placebo (saline solution that contained no drug). Periodically during the study, participants indicated the severity of the pain they felt and how much the pain was bothering them. When patients were on the IV with the placebo, both ratings were about the same. But when the IV was dosing morphine, a fascinating thing happened: The amount of pain they experienced remained the same, but the pain no longer bothered them as much. They felt the pain but were not in agony.

All of this is to say that what we call pain is more than a physical sensation, and the "feeling" part of pain isn't exactly proportional to the intensity of the signals emanating from our pain receptors. The effectiveness of many placebos is explained in part because signals indicating physical pain have to be interpreted by the brain. It turns out that if you believe you are engaging in an activity that will make you feel better, then you often start to feel better, even if there are no active ingredients in the treatment you're receiving. If you start doing something that has healed you in the past, then your brain has no need to keep reminding you that something is wrong, so it diminishes or eliminates the pain signal.

You often witness examples of this effect when you take pain relievers like ibuprofen. Many people end up experiencing relief from the pain faster than the chemistry of the drug actually works. For example, it generally takes twenty to thirty minutes for ibuprofen to dissolve in the stomach, enter the bloodstream, and physically affect the pain. But many people start feeling better in a few minutes. That happens because your brain recognizes that you have done something to ease the pain—the cavalry is on the way—and so it dulls the signal of pain.

The funny thing about placebos is that people are prone to discount their effects because they are "just in your head." But studies have demonstrated that placebo effects happen in part because the brain actually starts to release chemicals that dull pain in response to an *expectation* of relief, regardless of whether you are taking a real drug or a placebo. In creating an expectation of relief that translates into an *actual sensation* of relief, placebos act like real drugs.

WHAT ABOUT THE PAIN THAT ALL THOSE SINGERS ARE CROONING AND WAILing about? Emotional pain in story and song is most often described using metaphors and we use these conventions all the time without even realizing it. It is not uncommon for people to talk about the *weight* of responsibilities, the *lightness* of being, intellectual challenges that are steep *climbs*, feeling on *top* of the world, hitting rock *bottom*, *brittle* egos, and *iron* wills. Emotion has a direction to it: You feel *down* for a while, and then things start looking *up*!

Maybe the pain of love is just a metaphor that doesn't really involve physical pain at all. Or does it?

In another study, participants who had gone through a difficult breakup were asked to think about and reflect on what they had experienced. Some of them were given Tylenol® at the start of the experiment, and others were given a placebo. The people who got the Tylenol reported that they felt less bad about their breakup than did those who got a placebo, which suggests that feelings about the breakup involved real physical pain. We should add that other studies suggest that Tylenol actually blunts both positive and negative feelings, and it may be that Tylenol works just because it makes you feel less of everything.

Regardless, results like this teach us that heartache is, indeed, a real pain.

HEARTACHE

IS A

REAL

PAIN.

7

Do schools teach
the way children learn?

M OST OF US HUMANS REQUIRE AS MANY AS FIFTEEN TO TWENTY YEARS
of training before we're ready to live on our own and contribute to society, and culture provides structures that allow us to learn about and successfully navigate the environments
we are likely to encounter. For example, when babies are born, they
do not recognize that heights are dangerous, perhaps because infants don't yet move around on their own. Given that infants are
generally carried around by other people (and more than a few feet
off the ground), an innate fear of heights would lead most infants
to live in a state of terror much of the time. Playful parents hoist
babies into the air every day, eliciting squeals of delight from their
children. Heights? No problem.

But as babies learn to crawl, their world becomes a different
place. Now it's important to learn that heights are potentially dangerous. How do intrepid infants come to recognize the danger in a
flight of stairs?

A classic test of the ability to recognize heights employs a
device called the *visual cliff*, a two-tiered structure with a ledge

and a drop-off of several feet that is covered by a sheet of glass. (Skyscrapers like the Willis Tower in Chicago, and even the Grand Canyon, now have visual cliffs for adults!) When infants first learn to crawl, they confidently move themselves over the cliff and onto the glass, but within months *after* learning to crawl, they seem to figure out the potential danger. They crawl to the edge, look over, and back away.

When infants get an inkling that heights could be a problem, they look to adults, in a process known as *social referencing*, for a signal that will help them figure out what to do. In one set of studies with the visual cliff, moms were seated at the far end of the device (past the drop), and their babies were placed on the upper tier and allowed to crawl around. When they got to the ledge, they looked at their moms for guidance. Some moms were instructed to show fear as the infant approached the visual cliff. Their babies backed away from the ledge. Other moms were instructed to smile and show encouragement, and their babies continued to crawl past the ledge and onto the glass.

IN WESTERN SOCIETIES, PARENTS AND OTHER ADULTS TEND TO INTERACT with babies and toddlers directly, and children relish the attention. It's interesting to note that this behavior is not universal. In cultures where adults engage in less child-directed behavior, children learn mostly from observation and from their interactions with older children. In these cultures, children may watch an adult perform a task and then perform a version of it on their own (perhaps with toys).

The notion that all members of a society should possess a shared base of knowledge, and that this knowledge should be conveyed not only by parents, siblings, and peers, but also by institutions of learning, led to the development of schools—which are, in a sense, learning factories. The success of Western industrialization furthered the perception that we could systematize learning and instilled the confidence that we could scale up teaching, serving large groups of children, learning together in classrooms, school buildings, and grade levels.

In every factory, efficiency is a priority. And if we think of children as empty vessels to be filled with knowledge and skills, then the structure of typical schools seems like a pretty good idea. But children aren't empty vessels who simply need some knowledge and skills poured in; they're curious creatures with ideas, experiences, emotions, aspirations, and physical bodies that move—all of which play a role in what they remember and what they learn.

Traditional models of schooling are based on the assumption that the best way to fill those ready brains is by talking and showing and explaining, employing all kinds of media—books, lectures, films, and recordings—to provide information. But attempts to maximize efficiency in conveying content to large groups of learners often omit the essential components of learning that engage children's natural enthusiasm for discovery. Having children sit in rows and listen as wise adults explain things to them ignores two important aspects of young human beings: They're social, and they move. When most of the interactions in classrooms are between students and the teacher, students miss out on all there is to learn from interacting with one another. What might at first seem like pointless play is actually the process through which all of us learn deeply. Remembering and reciting are not sufficient. Social interactions among learners provide occasions for productive, collaborative, motivating muddling. Though the messiness of these kinds of interactions seems inefficient, it is essential.

Sitting quietly and still for long stretches of time is not a natural childhood activity. In fact, many teachers spend much of their time helping children overcome their inherent desire to move and do things. An orderly classroom is understandably appealing, but it inhibits crucial components of effective learning. It seems odd that our current concern over the sedentary nature of childhood in America is juxtaposed with school days dominated by sitting, with intermittent physical activity in PE and at recess defined as "breaks" from the hard work of learning. Physical activity should be *central* to the hard work of learning.

But knowledge doesn't live only in the head. The field of *embodied cognition* teaches us that we understand most concepts, at least in part, by interacting with them. In science, it is valuable for students to play with a pendulum, changing its length and weight to see how that affects its movement. And math instruction has a long history of using blocks and other manipulatives to help children learn place value, which can be quite helpful when teachers are clear about making the connection between the child's activity and the key math concepts being presented. This isn't a new idea: John Dewey, Alfred North Whitehead, and others were harping about the importance of physical movement for conceptual learning over a century ago.

WE COULDN'T CONCLUDE THIS ESSAY WITHOUT TALKING ABOUT TESTS—A near-ubiquitous feature of formal education and a topic of ongoing public debate. There are many good reasons to evaluate what learners know and what they can do with what they know. And research suggests that testing really does aid learning, both because students have to study for tests and also because the test itself helps the brain to recognize that information it encountered before will be needed again.

That said, if avoiding the threat of doing poorly on tests is the primary motivator for learning, we've got a problem. Many kids begin to feel as if the whole point of school is to remember things long enough to do well on the next test.

Many kids who are deemed "unmotivated" in school will spend hours playing video games, learning the guitar, or honing their ability to rap. It's not that they lack motivation; they simply lack motivation to do what the schools are asking them to do.

Leaders in education have begun to move away from the sit-and-listen model and toward the move-and-do model of learning, not because it's becoming trendy but because of its effectiveness in achieving the goals that schools are designed to satisfy: developing the minds of curious, engaged, happy learners.

Both of us play instruments, and we know that a key to our motivation to practice is the moment-to-moment evidence of our progress (or lack of progress). The same is true for any motivated learner who's learning to ride a bike, build a sand castle, or swing a baseball bat. Recognizing the relationship between the effort you expend and the return on that investment is a powerful motivator and a recipe for experiencing the joy of accomplishment.

When your only motivator is avoiding failure on a test, then the best feeling you can experience is one of relief ("Whew!"). Struggling with surmountable challenges and knotty problems and then conquering them leads to a different kind of emotion ("Wow!") that propels you forward to the next challenge.When kids have frequent opportunities to demonstrate to *themselves* what they know and how the things they have learned allow them to meet challenges and solve problems, they begin to appreciate that their learning matters beyond gold stars and good grades. People are natural learners, but not all learning environments capitalize on our human potential. In some ways, we need to make our schools less efficient in order to make them more effective.

REAL LEARNING IS A MESS.

Embrace the muddle.

Why do tongue twisters work?

She sells seashells by the seashore.

Peter Piper picked a peck of pickled peppers.

I saw Susie sitting in a shoeshine shop.

He slit a sheet, a sheet he slit, upon the slitted sheet he sits.

THIS LAST ONE WAS POPULAR WHEN ART WAS IN ABOUT SEVENTH grade, for reasons that will be obvious after you try to say it aloud three times fast.

When you do a radio show, you become very attuned to speech errors, and it's amazing how often people intend to say one thing and end up saying another. Speech errors come in different forms, and understanding the various types of speech errors helps explain how those tongue twisters work.

The individual speech sounds that make up language are called *phonemes*. The remarkable thing about languages is that each one uses relatively few phonemes to make up its entire vocabulary. Each unique word comprises a particular combination of phonemes.

Phonemes can be divided into consonants and vowels. Vowel sounds (*a, e, i, o, u*) are produced by the vibration of the vocal cords and are distinguished from one another based on the shape of the mouth, lips, and throat, which affects the way they resonate in the head.

Consonants sounds are created by closing or changing the shape of the vocal tract using the throat, teeth, lips, and tongue. For example, the *s* sound is produced by placing the tongue near the roof of the mouth and forcing air over the tongue. The *sh* sound also uses hissing air, but the lips are forward and the middle of the tongue is moved toward the top of the mouth to produce a different shape and airflow.

These individual phonemes combine to form syllables, many of which are consonant-vowel-consonant clusters (sometimes abbreviated as CVC clusters), like *sat*, *sag*, and *cat*. Any syllable has to have at least a vowel in it (so the words *a*, *I*, and *oh* are all syllables). Some syllables have a vowel and just one consonant (like *be* or *at*), but many have a vowel in the middle and consonants at the start and the end.

The initial consonant of a CVC syllable is called the *onset*, and the vowel and consonant at the end are called the *rime*, which (as Bob likes to point out) rhymes with *rhyme*. The way syllables are set up psychologically, it might be more accurate to use the acronym C-VC, because the onset (C) is less strongly connected to the rest of the syllable than the phonemes in the rime (VC) are connected to each other. The more tenuous connection between the onset and the rime explains in part why we really enjoy rhyming words (like *sat* and *cat*) and don't feel as strongly about words that share only the first consonant and vowel (like *sat* and *sag*). Most speech errors affect the onsets of syllables.

Sometimes we mistakenly anticipate a speech sound we need for the next word (saying *dad dog* instead of *bad dog*). Sometimes we persist in a speech sound from the previous word (saying *bad bog* instead of *bad dog*). Sometimes we swap the sounds from two sylla-bles (saying *dad bog*). These swaps are common enough that they've been dubbed Spoonerisms, after the Reverend William Archibald Spooner of Oxford, who reportedly made these kinds of errors all the time.

There are many factors that can cause these kinds of mistakes in speech. The faster you speak, the less time the brain has to plan the movements of your mouth, tongue, and teeth. Using lots of similar speech sounds in a single sentence can make it especially difficult to plan which sounds belong in which location.

Sentences in tongue twisters are designed to create a particular kind of speech error in which sounds get swapped between adja-cent words. The alternating speech sounds are placed at the begin-nings of words to increase the likelihood that the sounds will de-tach from the syllables where they belong and move to neighboring syllables. In "She sells seashells by the seashore," *she sells*, *seashells*, and *seashore* share a similar structure, but the onsets are swapped between the two syllables in each phrase. This verbal challenge is made more difficult because the positions of the mouth required to produce *s* and *sh* sounds are similar as well.

THERE ARE OTHER KINDS OF SPEECH ERRORS THAT ARE MORE ELABORATE THAN those caused by tongue twisters. Many of us have experienced a momentary difficulty remembering names of loved ones. It's not unusual for a parent to stand at the base of the stairs to call one of the kids and cycle through the names of *all* of her kids (and perhaps the family pet) before getting to the right one.

Names are a special case. Most words that we use apply to cat-egories of things. The word *cat* refers to a whole class of objects, but *Felix* refers to a particular individual. You find it harder to recall names in part because names refer to only particular people, and there is no obvious reason that people should have the name they have been assigned. People you know are grouped in categories (like the people who live in your house), so you often clump their names together in memory, which explains why you sometimes confuse the names of your children or other family members.

Another kind of speech error involves misusing a word because you don't know its definition. These mistakes are called *malapropisms*, named for the character Mrs. Malaprop from an eighteenth-century play by Richard Sheridan. In a malapropism, someone might say, "I thought he did a perfectly *superfluous* job," when intending to compliment someone, not realizing that *superfluous* means "unnecessary." (Bob frequently thinks that Art does a perfectly "superfluous" job on the show.)

Another kind of error involves simply retrieving the wrong word, mistakes that are often referred to as *Freudian slips*, under the assumption that selecting a particular word reflects a subconscious intention to say the word. As the old joke goes, "My therapist said to me, 'If it's not one thing it's your mother.'"

Freud proposed an elaborate system of subconscious desires— which, by the way, has little empirical evidence to support it. But if you substitute one word for another in a sentence, the word you misuse has to have gotten into your working memory somehow. One possibility is that it was suggested by the context. If you are distractedly watching trains at a station, you might insert the word *train* in a sentence because you're thinking about trains. If you're ruminating on a thought that is plaguing you, it's possible that you will say a word that relates to the thought, even in a sentence where it doesn't belong. But not every speech error is a reflection of some deep and hidden thought.

Finally, it is worth pointing out that most of the speech errors you make are not mistakes regarding the pronunciation or the definition of a word, but with the structure of the sentence itself. Our thoughts evolve constantly, and sometimes you switch what you intend to say midstream.

Written language is usually pretty orderly, because it is a product of a lot of thought and editing. In live speech, however, these edits happen on the fly. The funny thing is that you often don't notice how many stops and starts there are in sentences people speak aloud. Someone begins a sentence, pauses, and repeats a word a few times. Halfway through the sentence, there is a stop and restart. Of course, as you're listening, you are attending to the *meaning* of what is spoken more than you are attending to the words and sentence structure, so you rarely notice how disjointed many conversations are.

If you record yourself speaking, as radio hosts do all the time, you get to hear all the ways in which spoken sentences stop and start. If you consider yourself a skillful speaker, it may be painful to hear the errors that contaminate your speech. But life is not scripted with our lines edited for clarity and grammatical fidelity.

TO *err*
IN SPEECH
IS *human.*

A GOOD
tongue twister
IS *divine*.

Do we get more done
when we multitask?

CHANCES ARE, YOU'RE ONE OF THOSE PEOPLE WHO THINKS THAT MULTI-
tasking is a critical skill for the modern world. You're way
too busy to do just one thing at a time; it would be impossible to get everything done without juggling several tasks at once.
Not to mention incoming emails, texts, instant mess—

Hey, somebody just posted the cutest cat video on Facebook®!
Awwwww.

Sorry, where were we?

Oh yeah. Multitasking.

Chances are, you're one of those people who think that multitasking is a critical skill for the modern world . . .

Wait. We already said that.

Okay, what just happened?

You just read a little demonstration of how well the two of us
multitask, which is not well at all.

You might assume that when you multitask, your brain just
splits up its capacity among the various things you're trying to do.

If you surf the Web while talking on the phone, then maybe 40 percent of your brain is checking the Web and 60 percent is taking care of the conversation. (Or some other percentages depending on who's on the phone and what's on the screen.)

Is that really how multitasking works?

In a word, no.

You probably realize already that your eyes can focus in only one place at a time. People who try to text and drive at the same time (in other words, insane people) can't look at their phones while they look at the road, and they're not looking at the road while they're looking at their phones. But this idea extends beyond where your eyes are pointed. We've said before that we see with our brains, not with our eyes. Our eyes are simply light detectors that send signals to the visual cortex, which is what actually does the seeing.

In fact, if you are looking at something related to a task you're working on, and some new information comes into view that's related to some other task that you'd also like to be working on at the same time, the new stuff might escape your notice entirely. You can actually point your eyes right at it and still not see it. You can't really listen to more than one thing at a time, either. Just like seeing, we don't hear with our ears; we hear with our brains.

In classic experiments about attention, participants wear headphones that play a different signal to each ear. The sound in the left ear might be a passage read from a book, while the sound in the right ear might be a list of seven words repeated over and over. While they listen to the sounds coming into both ears, participants are asked to repeat aloud what they are hearing in the *left* ear, which of course requires that they pay extra attention to what is coming into the left ear.

At the end of studies like this, very few people can remember the words that were playing in their right ears. Their brains were simply too busy trying to focus on the input from the left ear, and focusing on one audio stream made it almost impossible to hear what was going on in the other one.

What may come as a surprise is that your mind almost never truly multitasks, if we define *multitasking* as "paying attention to two or more things at exactly the same time." What your mind *can* do is shift attention fast enough that it seems like you're paying attention to two or more things at once, even though you're really attending to only one thing at a time and rapidly switching your focus of attention from one thing to the other.

Brains evolved to help us perceive the world accurately and act on it effectively. The limitations of your sensory and action systems require that your brain prioritize which aspects of thought are going to determine where you move your eyes, which sounds you pay attention to, and how you activate your body to do things.

A set of brain mechanisms referred to as the *executive system* selects which task will be the focus of attention from among the many things in your environment. This is how your brain prepares to engage your sensory and movement systems to accomplish whatever it is you're working on. When you try to multitask, you force your executive system to shift attention from one task to another.

UNFORTUNATELY, SHIFTING BETWEEN TASKS IS MUCH LESS EFFICIENT THAN focusing on one task, completing it, and then moving on to the next one. And in situations in which at least one of the tasks is ongoing (time-dependent), like paying attention in a meeting or a class, shifting attention is especially problematic. It's certainly possible for you to check your email during a meeting (particularly if it's a boring meeting), but when you focus your attention to read an email or send a reply, whatever happened in the meeting while you were attending to email escaped your notice entirely. You missed it.

Of course, many of us try to "make good use of our time" while sitting in meetings, and because most meetings are slow and inefficient, we seldom feel the negative consequences of doing so. The problem is, you often don't know in advance when something important is going to come up, so if you happen to check out during the two minutes of the meeting when something actually important happened, you're out of luck.

There are problems with multitasking *even if* none of the tasks you're switching among require ongoing attention, like a conversation or a meeting or driving, because whenever you shift your attention from one task to another, there are *switching costs*. When you leave one task, however briefly, to attend to another, doing so requires that you remove information from working memory and orient to the just-switched-to task. When you return to the first task, you again have to reorient (replace the contents of working memory), recalling what you were doing when you were last paying attention. The reorientations required in leaving and returning to various tasks consume time, effort, and energy, and they also diminish accuracy and efficiency. It may feel to you like you're doing a good job of juggling, but chances are you're actually wasting time. Of course, the lack of efficiency is exacerbated by the increased likelihood of making mistakes.

It's certainly possible to *do* two or more things at the same time. What's not possible is *focusing your attention* on two or more things at the same time.

Nearly all parents have had the frustrating experience of trying to make contact with children who are immersed in a TV show or video game, seemingly beyond the event horizon of the black hole of technology. We know they can hear us. How can they just ignore us like that? Well, consider for a moment that in order to willfully ignore something, you actually have to perceive it. Most kids who are in the midst of trying to save the galaxy using only their thumbs are not ignoring their parents—they literally can't hear their parents.

If it's impossible to pay attention to more than one thing at a time, then why don't we recognize it? Most people continue to multitask at work because they assume that it is making them more productive than they would be if they did only one thing at a time. Many people swear by it.

Well, it just so happens that the brain mechanisms that allow you to shift attention from one task to another are the very same mechanisms that you use to monitor your own performance. In a sense, paying attention to how well you're doing is a form of multitasking—you are shifting back and forth between the attention required to do a task and the attention required to assess how well you're doing—which makes you not only a terrible multitasker, but terrible at evaluating your own "multitasking prowess."

ALMOST EVERY TIME WE TALK TO GROUPS ABOUT MULTITASKING, SOMEONE tells us they've heard that women are good multitaskers, which is usually followed by a story about how in our evolutionary history, women had to juggle several different tasks at once.

There was one study that showed a gender difference in which the women were less bad at multitasking than the men. There are two important things to say about this, though. First, even in this study, the men and women both got worse at both things they were asked to do when they multitasked; the women just got "less worse." Second, this gender difference hasn't held up that strongly in further investigations, so we're inclined to think that if there are any differences in multitasking ability between men and women, they are too small to be significant in practical situations.

Interestingly, there are a few people who actually seem to be pretty good multitaskers. Studies suggest that about 10 percent of people can do two complex tasks at once without suffering too much. That means that there is a 90 percent chance that you're not one of them. We don't really understand why these lucky (?) few are so endowed.

As Art is fond of pointing out, though, as bad as we are at multitasking, it is a wonder we can flip back and forth among tasks without completely falling apart. This ability is important when we get interrupted (as happens so often). If you are in the middle of writing something and your phone rings, you can take the call and still manage to get back to what you were writing. Without the ability to quickly recover what you were doing, each time you got interrupted, you might move on to do a completely new thing. (Sort of like Art's dogs, who are easily distracted from chewing on the carpet and then go on to find something else to chew on.)

In fact, habits are the one way that your brain allows you to do more than one thing at a time, precisely because habits, by definition, are behaviors that you can carry out without having to focus any of your precious executive resources to guide the flow of attention.

The reason you spend all of that time learning to touch-type, for example, is that once typing becomes a habit, you can type words and sentences without having to think about which keys you have to press to type each letter. Typing is a great example of how habits facilitate accomplishing goals. As habits become strengthened (more highly *automatized*), the actual details of your movements are no longer easily accessible to you.

Answer this question quickly, and without looking at a computer keyboard: Which key is to the right of the Y key?

See? If you're like most people, the answer to that seemingly simple question is not readily accessible, but you can type the letter U without any trouble at all. Your right index finger "knows" exactly where it is. Especially if you type really fast, trying to think about which finger is assigned to each key will not only slow you down tremendously, but will cause you to make mistakes as well.

Once you can type without having to pay attention to what your fingers are actually doing, you can devote your energies to thinking about the messages you're typing instead of how to operate the keyboard. This is true of all of the habits you execute throughout each waking day, and there are lots of them. Habits are a necessary part of human functioning. Human beings simply couldn't survive if we had to devote conscious attention to everything we do, so brains develop a nifty trick of operating our bodies automatically when it's possible to do so.

In closing, we want to emphasize that if you are going to perform at your peak on any particular task you're doing, then try single-tasking. You'll be amazed at how quickly and effectively you can get things done when you stop trying to do everything at once.

Multitasking:

DOING

LESS

by

DOING

MORE.

Can we be conscientious and creative?

HUMAN BEINGS ARE CAPABLE OF SOME PRETTY SOPHISTICATED THINK-
ing, which is one of the distinguishing features of our species. We can figure out solutions to all kinds of problems
and challenges—even those that are unlike anything we've encountered before. Functioning effectively in the world requires the
capacity to size up the situations we find ourselves in and figure
out what to do next. But is this how scientists think when they're
doing science? Or is scientific thinking different from what most
people do when they're navigating their lives day to day?

It turns out that what we call scientific thinking is *unlike* the
kind of thinking that most people engage in most of the time. Of
course, this doesn't mean that, once people become scientists, they
no longer think in unscientific ways. We both have lots of evidence
for this. In fact, some of the most unreasonable people we know are
scientists who think very effectively when they're doing science,

but seemingly abandon all of that training when they're sitting in a faculty meeting.

A fundamental difference between scientific thinking and "typical" thinking has to do with the overall goal. Science is all about *disconfirmation*, which means that science is focused on finding ways to show that a particular theory is *not* true.

Most people's everyday thinking, though, is all about *confirmation*. That is, when you look out at the world, you tend to focus on information that is consistent with what you already believe. You are much more likely to pay attention to information that fits with your existing beliefs than you are to focus on information that would require you to revise your beliefs. Indeed, this is such a pervasive part of the way people think that it has been given a name: *confirmation bias*.

Here is an experimental setup that illustrates this kind of thinking. Participants are told that there is a simple rule that describes a sequence of three numbers. Their job is to figure out the rule. Participants can test the hypothetical rules they come up with by suggesting other three-number sequences, and the experimenter tells them whether their sequences fit the rule.

To get things started, participants are told that the sequence 1, 3, 5 follows the rule. Most people start with a guess that the rule is "increase each number by 2." They then test their rule by trying some other sequences like 7, 9, 11 or even 51, 53, 55. Each time they try out another sequence of three numbers like this, they are told that their test sequence follows the rule. After several tests of three-number sequences that increase by 2, most people stop and state quite confidently that the rule is "increase each number by 2," which was confirmed in all of the sequences they tested.

But in fact, the actual rule (the one that the experimenter has in mind) is more general than that. The actual rule is "each number is larger than the one before it," so *any* increasing set of numbers follows the rule, including 7, 692, 1,400, and –7, 0, 7. Most people never test sequences that increase by amounts other than 2, let alone test decreasing sequences, and thus they fail to discover the more general rule. If someone were to test *just one sequence* that didn't increase by 2, like 7, 8, 9, the increase-by-2 hypothesis would have to be abandoned and a new hypothesis would be needed. That might clue them in that they need to test lots of other sequences as well.

The takeaway here is that participants in experiments like this *tend to focus only on tests that will confirm what they already believe.* They seldom test other circumstances or sequences in which their working hypotheses might fail. Again, this is how most people, including most scientists outside the lab, behave most of the time.

Science doesn't work that way. In science, if you want to test a conjecture or a hypothesis, you have to create tests that provide every opportunity for your hypothesis *not* to be true. In other words, you have to imagine all the ways that your pet theory may not be right, and subject your theory to those possibilities. And according to the rules of science, even if you find no evidence that disconfirms your hypothesis, the most you can say with confidence is that there is, as yet, no evidence that this hypothesis is *not* true.

We realize that this kind of language drives many nonscientists crazy, but it's a necessary part of the scientific method and a central component of scientific progress. There is a critical difference between saying that some proposition is "true" and saying "there is no evidence that it is not true." The first statement leaves no room for further investigation or refinement in the future; the second is tentative, and reflects the fact that you haven't tested your idea in every conceivable (or inconceivable) circumstance, so there remains a possibility, however remote, that it is wrong.

All this is to say that science provides a way to minimize the influence of confirmation bias. In addition to going out of their way to look for information that will disconfirm cherished beliefs, scientists agree to be governed by data. No matter how strongly a scientist wants a particular theory to be true, if the data come out differently, the scientist is forced to change his or her beliefs. In this way, the rules of science are designed expressly to help people think more effectively than they might otherwise.

WE DON'T WANT TO GIVE THE IMPRESSION THAT SCIENTIFIC THINKING IS always good and nonscientific thinking is always silly. It's not that simple, even though the term *confirmation bias* suggests it is generally a bad thing. Otherwise, we would call it something more positive, like confirmation tendency or confirmation-based reasoning.

In fact, confirmation bias is a pretty good idea most of the time. For example, suppose Art has the belief that he should not tell dirty jokes in class. It would be a good idea for Art to let his actions be guided by that belief. It is possible, of course, that he is wrong and that it is fine for him to tell dirty jokes in class, but the only way to find out for sure would be to tell some dirty jokes in class—to put his hypothesis to the test. If he was right all along, then he could get himself in a lot of trouble—all in the name of good scientific thinking. So Art is better off sticking with actions that continue to confirm his existing belief.

In many real-world situations, the consequences of testing a belief by trying to disconfirm it can be big and bad. We are better off missing out on a few opportunities than suffering the consequences of trying something that fails spectacularly.

Not everyone follows rules so closely, though. Bob is more likely than Art to break a few rules just to see what happens. Not that Bob is looking for all-out anarchy, but rules and deadlines are just not as important to him as they are to Art.

It turns out that there is a basic personality characteristic that relates to this difference. One of the Big Five personality character-istics is *conscientiousness*, which reflects how much people like to complete the tasks they start. The more conscientious you are, the more you're motivated to complete tasks and follow rules (includ-ing rules for yourself that you come up with on your own, like not telling dirty jokes in class).

The world tends to reward conscientious people. In companies, managers tend to notice the conscientious people because they are the ones who can be relied on to finish the assignments they are given. They focus. They follow the instructions to the letter. As a result, they are often chosen for promotions and rise through the ranks to positions of greater responsibility.

Conscientiousness seems like a personality characteristic in which there is only one good end of the continuum. Being highly conscientious looks like a great thing, while being low in consci-entiousness seems like a recipe for disaster. People who don't get things done on time, who need to be reminded repeatedly to finish work, and who treat the rules as mere guidelines would seem to be at a disadvantage in life.

Yet there must be *some* advantage to being low in conscien-tiousness. Over generations of human evolution, the pressures of natural selection have likely ensured that any characteristic that is always bad does not survive. So if there is a range of conscien-tiousness in the human species, there must be an advantage to low conscientiousness, at least some of the time.

And there is.

If you are too highly conscientious, then you spend your en-tire life following rules. You color inside the lines and believe that strictly adhering to the rules is the right way to do things. Always.

And that makes you less creative.

CREATIVITY REQUIRES BREAKING RULES. PEOPLE WHO ARE THE MOST CREATIVE often break rules in their domain of expertise. Not all the rules, necessarily, but some of the rules. In the rise of abstract painting, for example, Braque and Picasso (following the lead of earlier painters like Cézanne) worked to represent multiple views of the same image in a single painting. Prior to Braque and Picasso, though, artists would create subtle variations in viewpoint within a painting so that it still generally appeared to be a coherent work.

The cubist style that Braque and Picasso created was like nothing that had been attempted before. It required ignoring many of the conventions of figural painting to create images that were in some instances virtually unrecognizable at first glance. Not every viewing audience appreciated their creations, but Braque and Picasso persisted in trying new styles.

Similarly, when Einstein was working through his special theory of relativity, he had to break some of the prevailing fundamental rules of physics, including the assumption that mass, space, and time were fixed values. Rather than assuming his theory must be wrong if it implied that the velocity of an object affected its mass (which seems intuitively weird), he pushed ahead, and in the process he changed the way physicists think about the universe.

What's interesting about these creative individuals is not only that they broke the rules within their domains of expertise, but they also tended to flout social conventions. Creative people often treat rules lightly, as guidelines to be followed when convenient rather than as impenetrable boundaries in life. Highly creative people often live their lives at the fringe of what is socially acceptable.

There are so many stories of artists and musicians who ignore the rules of typical behavior that it has become a common stereotype. But the same thing holds true for many other creative people. For example, the autobiographical books written by Nobel

Prize–winning physicist Richard Feynman illustrate quite vividly that he cared very little for what was expected of him by others in social situations.

In fact, it is true that the more you tend to be driven to follow rules and meet deadlines, the less likely you are to be creative, so here's a special homage to Bob and people like him:

PUNCTUAL

CREATIVITY

is an

OXYMORON.

Is it true
that we only use
10 percent of our brains?

NOT LONG AGO, WE WERE DISCUSSING THE REACTIONS WE GET WHEN people learn that we're psychologists. Some people look a little sheepish and mutter something like, "I bet you're analyzing me now," as if we've got some kind of magic psycho-mojo. Art generally responds that people shouldn't worry, because he doesn't really care about their problems. Bob lets people know he does care about their problems, just not professionally.

After our new acquaintances get comfortable with the fact that we're not dissecting their psyches, they often ask us questions about how people think. One of the most frequent questions we hear involves some variant on the claim that human beings use only 10 percent of our brains. Some people want to know whether it's true; others want to know how to prod more of their dormant brain parts into action.

The statement that we use only 10 percent of our brains has gained tremendous traction over the years. We just typed "10%

brain" into Google and got 253 million hits. Happily for us, though, most of the links that show up first include the word *myth*. But even with all the webpages' careful explanations of why it's not true that we use only 10 percent of our brains, this myth—like many myths—persists. (Say that three times, fast.)

Stories in popular media make great hay promoting the idea of untapped brainpower. In the 1991 movie *Defending Your Life*, one of the accomplishments that allowed souls of the deceased to progress to "the next level" (whatever that is) was the ability to harness some of that unused brain. In the 2014 movie *Lucy*, Scarlett Johansson gains superpowers after she ingests a drug that suddenly gets her brain running at full capacity.

The idea that most of our brains are sitting idle through much of our lives has become a convenient dramatic convention about the promise of hidden human potential. In reality, though, we all use all of our brains, all the time.

THE BRAIN IS AN EXTRAORDINARILY ENERGY-HUNGRY ORGAN, REPRESENTING about 3 percent of an average person's body weight while using 20 to 25 percent of a person's daily energy supply. That's right: Your brain is burning several hundred calories a day, every day, even when you're lying in a hammock and napping.

You might wonder why the brain uses so much energy. After all, it would seem that our muscles, doing all that walking, grasping, lifting, and stair climbing, ought to burn more calories than a brain that's just riding around comfortably inside the skull. But once you understand the work that the most active cells in your brain have to do, you begin to recognize why that three pounds of precious goo in your head requires all that energy. Cells called *neurons* are the fundamental workhorses of the nervous system. These very special cells, which live in the brains of flies and sea slugs and humans alike, form the electrochemical circuitry of the brain. Neurons function by sending electrical signals from the cell body down a long projection called an *axon* to the next neuron in a circuit. Some axons are relatively short, but others are as long as a yard. Their electrical signals are generated chemically by the motion of charged particles flowing in and out of the cell.

When an electrical signal travels down the length of the axon and reaches a terminal end, it causes some remarkable molecules called *neurotransmitters* to be released into the tiny space, called a *synapse*, between one cell and the next. These neurotransmitters work by attaching to tiny receptors on the adjacent neurons, telling them that a signal has arrived. Because each neuron can send signals to as many as ten thousand other neurons, there is an awful lot of chemical activity going on each time we take a breath, raise a hand, blink, or think. And that's just the neurons we're talking about. There are other cells in the brain called *glia* that perform additional structural and functional tasks, and that are at least as numerous as neurons.

It takes a lot of energy to make all this stuff happen, and the brain uses more or less the same amount of energy no matter how hard you're thinking. If you are really concentrating on something, the brain uses a bit more energy, but not much compared with what's required just to keep the lights on.

Evolution would not allow an organ to burn that much energy and then not use all of it, all the time. Indeed, the brain is a buzzing hive of activity. Even when you're asleep, your brain is active, continuing to process memories of events and skills from the waking day and doing some housekeeping chores related to chemicals and toxins that build up while you're awake.

So if you're actually using all of your brain, all the time, where does the 10 percent myth come from? There is no definitive single source for this myth, but it probably reflects two things.

First, as physiologists in the nineteenth century began to dissect brains and to understand how they're put together, they discovered that brains were composed of a lot more than just neurons. There are a number of fluid-filled chambers in the brain that provide nutrients and protection and allow the brain to float inside the head. And there are the glial cells that we mentioned earlier, performing their various support functions. If you assume that only the neurons in your brain are actually doing anything important (which isn't really true), then you could imagine someone saying that you are only really "using" 10 percent of your brain. But of course, that is misleading in a couple of ways.

First, all of the structure of the brain is crucial for making sure that it works effectively. If the brain weren't protected and couldn't get nutrients and clear away toxins, it just wouldn't work.

Second, when people hear that they are only using 10 percent of their brain, there is an immediate presumption that if they used more of their brain, they would be smarter or more effective (or in the case of the movie *Lucy*, they would have superpowers that would allow them to manipulate matter—untrue, but pretty cool).

Starting with the psychologist William James in the late 1800s, several authors who thought about the mind directed their attention to what people are capable of achieving. In addition to William James, the early-twentieth-century business guru Dale Carnegie, whose classic book *How to Win Friends and Influence People* was first published in 1931, tried to get readers to think about how much more they could achieve through hard work and learning.

Without drawing on scientific data, these writers assumed that the human capacity for learning new facts and skills was virtually limitless. They observed that highly successful people continue learning new things throughout their lives while still remembering much of what they learned as children. Consequently, it seemed as though most people barely scratched the surface of what they were able to accomplish.

It is hard to put an exact percentage on the amount of mental capacity that people use, because we don't really know the limits of what's possible, but all of the available scientific evidence suggests that the harder you work, the more you learn across the life span. And this view of thinking probably has a kernel of truth to it. Most people do not take advantage of all the opportunities around them. Many people shy away from learning things that do not seem directly relevant to what they are trying to accomplish at the moment. People often talk themselves out of learning new skills as adults. In that sense, most of us are capable of far more than we actually achieve.

So the next time someone asks you whether people use only 10 percent of their brain, you can tell them that we use all of it—but even so, the vast capacity of a mind is still a terrible thing to waste.

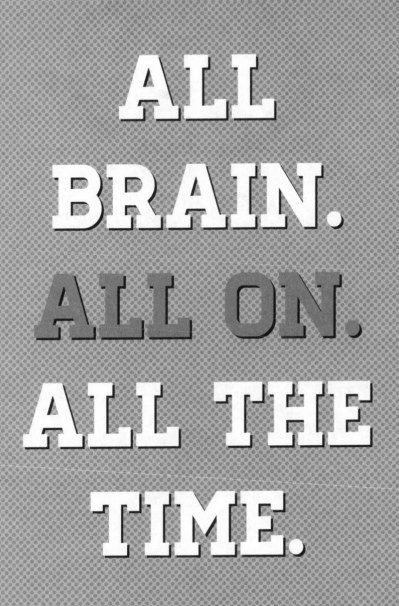

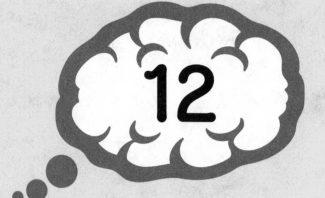

Is our memory doomed to fail?

WE SAW A T-SHIRT THE OTHER DAY THAT IDENTIFIED THE WEARER AS a sufferer of CRS, which was helpfully defined on the shirt as CAN'T REMEMBER S***. As baby boomers and Gen-Xers drift into the autumns of our lives, our thoughts turn increasingly, understandably, to problems with memory. But from the way many people talk about it, you'd think that your mental capacities are destined to completely fall apart before your very eyes. Is this true? Is this an inevitable part of aging?

Well, there's some bad news and some good news. Let's do that bad news first. Starting in your early twenties, you begin to undergo a long, slow cognitive decline. When you are in your early twenties, your cognitive capabilities are at their highest: You are able to think fast, and your memory works at its best. From then on, your speed of thought begins to slow, and it takes longer and longer to learn new things. (You *can* teach an old dog new tricks, but you and the dog have to be more patient.)

There's a lot of good news, too, though, which both of us old dogs are very happy about. Perhaps the best news about your long, slow cognitive decline is that it's long and slow, which means that well into your seventies and eighties, you may not really experience much of a problem in most aspects of thinking. The functioning of a healthy brain doesn't get that much worse just because it's getting older. Brains usually decline in their functioning because of disease, strokes, microstrokes, and brain injuries.

Another bit of good news is that in many cases it actually gets easier to learn new things as you get older, because the best way to learn is to attach new information to things you know already. Think about watching a World Cup soccer game, for example. People like the two of us, who know next to nothing about soccer, will remember very little about the game later. Perhaps we'll recall that a goal was scored and that that there were several moments when someone was writhing on the ground in agony until people stopped watching him. A real soccer fan, though, will remember all kinds of things about the game, including how great plays got set up, what led to the goals that were scored, and the controversy over a particular substitution late in the game. The knowledgeable viewer remembers a lot about what happened because information about the game can be integrated into her extensive base of knowledge.

MOST PEOPLE ARE FORTUNATE TO PICK UP A LOT OF EXPERTISE AS THEY GET older, and they can rely on their expertise to help them learn. As Bob likes to point out (with just the right amount of smug satisfaction), young people need to learn faster than older people because they're so much worse at it. Even with all that wonderful speed that younger people have, they still won't learn as much about a topic as someone who already has a fair amount of knowledge that can function as a scaffold for understanding and interpreting new information. All of this is to say that there is no reason to fear the mental consequences of getting older, because most of them are actually good.

We realize, however, that our telling you that memory doesn't get that much worse as you get older seems to fly in the face of many people's own experiences. Talk to people in their fifties, sixties, or seventies, and they'll tell you stories about actors' names they can't remember, keys they can't find, walking into rooms only to forget why they went there, and going to the store without a list and forgetting what they were supposed to buy.

How can we explain that? Well, a lot of the explanation is not really about what is wrong with your memory, but rather what is wrong with the way you think about your memory.

Art has three kids who are still smack in the prime of their cognitive lives. They haven't started that long, slow decline, yet they forget stuff all the time. Ask them why they forgot to do a homework assignment, and they'll say, "I spaced." Ask them why they forgot to empty the trash, and they'll reply, "Oh, oops." At no point after forgetting something crucial, though, did any of them say, "Oh no, I just had a 'Senior in High School Moment.'"

Yet it seems that after many people turn fifty, they begin to assume that each forgetful moment is a sign of an impending cognitive apocalypse. Each piece of information that cannot be accessed exactly when needed is interpreted as one more bit of evidence of mental doom.

This overinterpretation of intermittent forgetting not only is wrong but may actually be detrimental. Studies suggest that one of the worst things you can do for your memory, ironically, is to worry about your memory. Lots of research shows that it's harder to think and remember things when you're stressed. The amount of information you can keep in mind at any given moment gets smaller when you're stressed, and it's harder to think flexibly. And all of that affects what you're able to pull out of memory.

Research shows that even brief encounters with positive and negative information about memory and aging can affect adults' ability to remember. In several studies, older adults read either a brief article focusing on the fact that memory gets worse as you age or one indicating that memory doesn't really get that bad in older adults. Then they took a memory test. Sure enough, the group that read the more positive article didn't do that much worse than a control group of college students (who were still in the prime of their cognitive lives). The group that read the negative article suggesting that memory really *does* get worse performed badly on the memory test.

That said, there are some things you can do to help your memory as you get older. Most important, take care of your brain. A lot of the memory problems people experience later in life are the result of things they did early in life. If you enjoy a regular sleep schedule, and you avoid overdoing drugs, drinking too much alcohol, and sustaining sharp blows to the head, you're off to a good start.

One thing that happens to lots of people as they get older is that their sensory systems become less sensitive. All of that loud music you listened to as a kid causes hearing problems later in life. Cataracts and retinal issues can cause problems with your vision. Even your sense of smell can weaken. As your senses dull, the information that reaches into memory grows weaker as well, which can make it harder to retrieve information later. So it's important to get regular checkups to make sure that your eyes and ears function as well as possible for as long as possible.

Finally, make sure that you remain a lifelong learner. Lots of studies demonstrate that the more education you have and the more actively you *continue to learn* throughout your life, the longer it takes for signs of decline to appear. Education doesn't protect the brain from damage, but it does create lots of different ways to solve problems and remember information.

And remember that the belief that memory gets worse as you get older can become a self-fulfilling prophecy, so:

Forget the Senior Moment.

13

Why are continuity errors in movies difficult to catch?

LOTS OF PEOPLE SPEND LOTS OF TIME WANDERING AIMLESSLY AROUND the Internet. Students do it to avoid homework, and in the process manage to turn a forty-five-minute assignment into a four-hour monstrosity. At the office, people engage in fake work, randomly surfing websites rather than doing what they're supposed to be doing. And people avoid talking to one another when standing in line at the store by checking out their favorite websites.

One great website for aimless browsing is IMDb, the Internet Movie Database. If you're a movie fan and you browse IMDb, as Art is fond of doing late in the day at work, you might start by looking at the names of people who make movies: actors, directors, writers, and crew. Among the many factoids available on the website is a link to an enticing collection of mistakes called "goofs."

These goofs come in many flavors, but perhaps the most interesting of them are continuity errors. For example, there might be a shot of a character having a conversation while smoking a cigarette. The scene cuts to the other conversation partner and then back to the first person, whose cigarette is longer than it was before. Or a clock may move back and forth in time across shots. Glasses may disappear from a table and reappear in the next cut.

These kinds of continuity errors are easy to make because most movie scenes are assembled from many different takes. Despite the vigilance of set directors, prop masters, and script supervisors, small things may slip below their radar, and when the film gets to its final edit, those small discontinuities can work their way into the final product.

What is most amazing about these continuity errors, though, is how rarely you notice them. In fact, most of the time you watch a movie without recognizing any mistakes. Afterward, if you read about the movie, you might find several of these goofs that you didn't see when you first watched it. Watch the movie again, after knowing what to look for, and there they are, clear as day. Art read about a scene in the movie *Avatar* in which golf balls on the ground switch places across takes. He didn't notice that the first time he saw the movie, but he did notice it after reading about the error.

How is it possible that we miss these sometimes glaring inconsistencies?

WHENEVER YOU OPEN YOUR EYES, YOU HAVE AN IMMEDIATE AND POWERFUL conscious experience of the world around you—one that seems to contain a lot of detail about what is happening in your field of view. From this you might assume that your eye works like a camera, with entire scenes captured by some kind of light-sensitive apparatus, after which the brain figures out what exactly you're looking at.

But that's actually not what happens at all. Your sense of the visual world around you is constructed from lots of pieces. Light enters your eye through the *cornea* and is focused by a *lens* on the *retina*, a mat of light-sensitive cells in the back of your eye. Your retina is not like the light sensors at the back of a digital camera, which create an image that is equally high in quality across the entire image. Instead, there is a small area of densely packed cells in the middle of the retina called the *fovea*—the only part of the retina that creates a really high-resolution, well-focused image.

To see how small the fovea actually is, hold your arm straight out in front of you and focus your eyes on your thumbnail. Think about all the stuff that's not your thumb, and you will realize that everything other than your thumbnail is not sharply focused. You don't notice this most of the time because your brain is constructing an image of what you see. Recall that you don't see with your eyes or hear with your ears; you see and hear with your brain.

In order for the brain to construct a clear picture of what's in the world around you, your eyes are constantly in motion. They jump from place to place in order to gather information about what's in your field of view. Of course, you don't notice all of this jumping around because you suppress the input from the eye so that the world does not appear to be jiggling. Your eye *fixates* on a location, your brain takes in some information, then your eye jumps to a new location and fixates again (the jump is called a *saccade*).

Your visual system is able to get away with building up a sense of the world from a collection of short fixations, because your brain assumes that most objects in the world are going to remain where they are. A few things—like animals, people, or cars—tend to move, but they usually move in continuous and fairly predictable ways. Most other objects—like tables, trees, and teapots—generally stay in the same place unless someone or something moves them.

Objects that are clearly moving tend to attract your attention, because evolution has taught the visual system that moving objects are probably important for understanding what's going on around you. These objects might be potential threats or potential prey. A special kind of eye movement called *smooth pursuit* allows you to follow moving objects without having your eyes jump from one fixation location to another. You can see smooth pursuit in action by asking a friend to focus on your fingertip as you move your hand around in front of her and watching her eyes as they track your finger.

Because you assume that the objects that are not obviously moving will stay where they are, you don't store that much information about the visual world from one fixation to the next. Instead, your visual system assumes that if you need more information about a particular object, you can always look back at it later.

Your assumption that the world is pretty stable holds up almost all the time. Two big exceptions are magicians and movies.

MAGICIANS DO AN EXCELLENT JOB OF MAKING CHANGES TO THE ENVIRONMENT that you do not notice in order to create the illusion that objects have appeared or disappeared. A classic demonstration of this kind of trickery comes from a study by Dan Simons and Dan Levin. Here's the setup: An experimenter dressed like a construction worker approaches a person on the street and asks for directions to a location on a map. As the experimenter and the unwitting participant are talking, two other people carrying a large door walk between the two of them, blocking the participant's view of the experimenter dressed like a construction worker. As the door passes, the experimenter, who is momentarily out of view of the participants, changes places with one of the people carrying the door, who is also dressed like a construction worker. The "new" construction worker then continues the conversation with the person on the street.

You might guess that everyone who experienced this was shocked when the person they were talking to was suddenly replaced by someone else in the middle of a conversation, but that's not what happened. Only a small percentage of the participants noticed that they were talking to a different person after the door had passed. Because they were still talking to a construction worker, they didn't notice that the identity of the person had changed, even though the two experimenters looked and sounded different.

Failing to notice fairly significant changes in your visual field following a disruption is called *change blindness*. Carrying a door between people having a conversation is a disruption. A loud noise and flash (the kind magicians often use) is a disruption. And the cut from one scene of a movie to the next is also a disruption.

You miss most continuity errors in movies because, just like in the real world, you don't record that much information from one visual fixation to the next. This isn't a system malfunction, though; it's precisely the way brains evolved to work. As we write about in other chapters, brains are biologically expensive to run, and there is no real evolutionary advantage to remembering every detail about what we see, so we tend to remember the things that seem to matter (often, things that move) and very little of the rest.

In fact, the only way to really notice a continuity error in a movie is to fixate on the discontinuous detail right before the disruption and keep looking at that spot. Most of the time your fovea is pointing at what you're paying attention to, so paying attention to an object that disappears or changes its properties will lead you to notice the change.

Most movie editors will tell you that if they have to choose between a great take (when the actors nailed it) with a continuity error and a less-good take that preserves continuity, they will choose the take with the error because few people are likely to notice it.

The next time you read the list of goofs in a movie, don't feel bad that you missed most (or all) of them. Your visual system is working exactly the way it's supposed to. The only way to really catch goofs is to spend your time looking for them, which means you probably won't pay much attention to the rest of the movie.

We see
much less
than we think we see.

Are all
narcissists alike?

Aᴄᴄᴏʀᴅɪɴɢ ᴛᴏ Gʀᴇᴇᴋ ᴀɴᴅ Rᴏᴍᴀɴ ᴍʏᴛʜᴏʟᴏɢʏ, Nᴀʀᴄɪssᴜs ᴡᴀs ᴀ beautiful, vain, and arrogant young hunter. Nemesis, the spirit of revenge, was angry at Narcissus for spurning the mountain nymph Echo, who was only able to repeat the last words spoken to her (get it?) and who wasted away pining for Narcissus. Nemesis led Narcissus to a pond where Narcissus fell in love with the sight of his own reflection, and he, too, wasted away, not being able to tear himself from the beauty of his image reflected in the water.

His name lives on, not only as a genus of daffodils in the Amaryllis family, but also in our cultural lexicon as a term for self-absorbed people. The psychological definition of narcissism has gone through an interesting transition since its inception. At first it seemed as though narcissists were simply people who thought particularly highly of themselves, but it eventually became clear that many individuals who broadcast their profound levels of self-esteem are actually somewhat fragile and needy.

Think of narcissists as self-esteem vampires, feeding off the energy of other people. Not only do they need to hear an ongoing stream of accolades from the people around them, but—not being able to share the stage—they often actively diminish those same people.

Perhaps it's not surprising that there are lots of ways people can exhibit narcissism, and some of them have more dangerous side effects than others. For example, *grandiose* narcissists focus mostly on themselves, believing that other people should always pay attention to what they have to say.

Other narcissists are more focused on threats to their self-esteem that other people may represent. Particularly threatened by others' successes, these *vulnerable* narcissists are concerned with tearing other people down as a way of lifting themselves up, so to speak, in both their own minds and the minds of others.

Narcissists who are really good at what they do are quick to tout their own accomplishments, and because they have a finely honed set of strategies to gain attention from other people, they often obtain recognition for what they do. They seek leadership roles in order to be the center of attention and frequently convey an attitude that they are better than the other people in their social and professional spheres.

This combination of traits generally leads to attaining positions of high status in groups fairly quickly, although many narcissists (particularly the vulnerable kind) have a hard time holding on to their status and popularity.

Most people who aren't narcissists want to get along with other members of their group, so they behave in ways that demonstrate respect for others' strengths, skills, and opinions. They are kind to other people and consider others' feelings, needs, and desires.

Narcissists adopt a different strategy. Because they feel that other people should listen to them, they regularly force their opinions on others while criticizing the opinions of people who disagree with them—especially those they perceive to be status rivals.

Narcissists' strategies often pay off, at least initially. Others may be impressed by their confidence and perhaps swayed by the criticisms that narcissists level toward others. Over time, though, as nearly everyone in the group eventually becomes a target, group members begin to recognize that the criticism is not a reflection of informed understanding, but rather a way for the narcissist to assert superiority over rivals.

And although narcissists routinely find ways to get promoted to positions of authority in organizations by touting their accomplishments to superiors, groups led by narcissists usually suffer. Narcissistic leaders tend to take personal credit for group successes and shift the blame for problems to others. This way of taking credit and passing blame makes working for a narcissist downright maddening.

Many times romantic relationships with narcissists begin swimmingly. Narcissists enjoy the attention of new partners, of course, but their relationships often come to feel one-sided. They're also particularly hard to break off, because narcissists are threatened by the withdrawal of affection and attention. One of the most dangerous aspects of narcissism is *narcissistic rage*. Remember the vulnerable narcissists we mentioned earlier? When they sense other people withdrawing their support or the rising influence of a rival, they might respond to the threat by aggressively lashing out, yelling, or even resorting to physical violence.

WE SHOULD PROBABLY END THIS ESSAY WITH A REMINDER THAT NOT ALL BE-haviors associated with narcissism are all bad, all the time. Leading a group to work toward a goal, for example, requires a willingness to make yourself the center of attention. And it's important to take pride in your accomplishments and to let others know about your successes. It is not unusual for individual achievements to go unnoticed, and it's often valuable for employers, supervisors, and others in positions of authority to learn about what you do.

And yes, there is perhaps a bit of narcissism in everyone. Many people experience pangs of jealousy, for example, when friends or colleagues succeed or are recognized for their accomplishments. And it is tempting to minimize their accomplishments to boost your own feelings of self-worth.

Consider, though, that you can choose to behave in ways that thwart the little narcissistic demons that pop up from time to time. Rather than thinking of ways that your obviously undeserving colleague was unjustly rewarded, you can make a conscious effort to congratulate him on his good fortune. You might think this kind of behavior—offering congratulations based on conscious deliberation rather than a felt need—is insincere on your part because it doesn't reflect what you *really* feel. But acting in ways that reflect the feelings you would *prefer* to have is precisely how to evolve a more productive and positive mode of thinking. If your initial feeling is jealousy, but you consciously decide to act in a way that is generous and supportive, who's to say that your consciously initiated behavior is not the *real, sincere* you?

Most people who try taking action like we've just described find it remarkable how changing their habits of behavior begins to change their habits of thought. Rather than behaving like a jealous jerk, because that's what your feelings tell you to do, acting like a generous and gracious colleague actually "tells" your feelings to change. And indeed, most people are pretty good at reading their own narcissistic tendencies, which is why one of the most effective scales for determining people's level of narcissism is to ask them how narcissistic they are.

There are a few things you can do by way of self-preservation if you find yourself in the path of a narcissist. The first is to observe carefully and determine what kind of narcissist you're dealing with. Grandiose narcissists can be managed—just make sure that they eventually think every good idea was theirs. Give them the praise they want, and nod attentively at their stories. Vulnerable narcissists are a different story, though. Because they are prone to get angry and controlling when people disagree with them, it is better to avoid vulnerable narcissists than to engage with them. As frustrating as it may be to put your opinions aside and go along with a narcissist, it is not worth engaging in a battle you can't win.

Anyway, here's something to silkscreen on a T-shirt and give to the good-humored narcissist in your life:

But enough about me... What do YOU think about me?

Does time speed up as we get older?

WHEN ART WAS A KID, HE WAS LUCKY ENOUGH TO KNOW ALL FOUR of his grandparents. He got to see them on weekends and hear them tell stories about when they were young and what life was like when Art's parents were kids. Invariably, they would sigh wistfully at how fast the years had passed.

Time passing quickly? This didn't square at all with Art's experience as an eight-year-old. He felt like time moved slowly. It had been *ages* since he was six. Of course, now Art looks at his twenty-two-year-old sons and wonders how they grew up so fast.

Nearly everyone talks about time speeding up as they get older. Why is that?

Let's begin by explaining a little bit about perceptions of time. It turns out that there is a difference between how we experience time in the moment and how we remember it when we think about the past.

Most people recognize that time seems to go faster when you're completely engaged in what you're doing than it does when you're bored. The psychologist Mihaly Csikszentmihalyi (pronounced chick-SENT-me-hi) developed a concept he called *flow*—the experience of being so immersed in what you're doing that you lose track of time. People who play video games or engage in other tasks that require sustained attention often experience it. Great conversations, playing sports, and reading can also lead to a flow state.

Boredom is the opposite of flow. Think about the last time you were sitting in a waiting room at a doctor's office or in an airport or train station. You can't completely disengage from the environment, because you need to devote some attention to listening for your name to be called or for the announcement that it is your turn to board the plane. Your need to listen or watch for a signal to move prevents you from devoting all of your attention to something more interesting, like a book or a puzzle, even though waiting for a signal can be one of the most boring things imaginable. As a result, you start paying attention to the passage of time. You are aware of each minute, sometimes each second. Time moves painfully slowly, and you experience every last bit of it.

But when you look back on a period of time from the past—say, last week or last month—your perception of how long that time period "feels" is affected by the number of distinctive events that you can remember from the period you're thinking about. Consider what happens when you move to a new house or apartment. The days seem to fly by as you carry and unpack boxes, deciding where furniture belongs and where pictures will hang. Each day is filled with lots of brand-new experiences and events, and because you probably don't move that often, you remember all kinds of details about your new place and how to get around the neighborhood.

When you look back on the week you moved, it seems very long because you can remember many specific and unusual events—the unpacking, the new neighbors, the little discoveries (good and bad) that you often come upon when you move into a new place.

Because the week was packed with distinctive events that are now recallable, it feels psychologically like the week lasted a long time. But compare an eventful week like that with a typical week at work or at school. Most weeks follow established routines. You wake up each morning at about the same time. You get yourself washed and dressed in the normal way. You head out of the house and take the same route to work or school. The days move according to a normal rhythm, but when you look back on a typical week like that, you wonder where the time went. If there weren't a lot of distinctive landmarks, it is hard to remember many of the week's particulars. Each day's commute to work or school was pretty much like every other day's, so your brain doesn't store a lot of the details.

The routines of work or school can likewise blur together from one day to the next. Because your routine weeks include few interesting events to store and later recall, remembering those weeks makes them seem as though they passed relatively quickly.

As you get older, more of your life involves doing things you have done before, as opposed to things that are new to you. The transitions to adulthood involve the establishment of routines that make life manageable. Having children creates an even greater demand for schedules that balance the many unpredictable aspects of parenting with a sense of predictability. The first few months of a baby's life are likely to be very distinctive for new parents, as are the various milestones of maturation as children grow up. But routines are the stuff of a stable life.

When you repeat a set of actions in a given situation, you eventually learn associations between the situation, the actions, and the outcomes the actions produce. Those learned associations create habits, and once you develop a habit, you can perform the routine without having to devote much attention to it. Habits provide added efficiency to our operating in the world. But the activities that you perform by habit (with little thought or attention) tend not to stick in your memory, and you're unlikely to remember them later. When you look back and recall days in which you engaged in a lot of habits, those days will feel shorter than those days in which you did lots of new things that required your attention.

YOUNG PEOPLE'S LIVES ARE FILLED WITH MANY NEW EXPERIENCES PRECISELY because young people have not had much time to accumulate a lot of memories. Older people's lives include fewer new experiences precisely because older people who've been around the block a few times are repeating much of what they've experienced in the past. As you age, the increasing familiarity of repeated events over time makes individual instances less and less memorable. If there's less to remember about a given interval of time, then that interval in retrospect will seem to have passed quickly.

Another theory of time perception—one that's been around since the late 1800s—suggests that we experience time as a proportion of our life spans. It's not surprising, then, that eight-year-old Art thought his sixth birthday was ages ago, considering that those two years represent a quarter of an eight-year-old's life span. But two years represents only 4 percent of the life span of a fifty-year-old, and this difference has been proposed as one explanation for the apparent speeding up of time as we age.

The cycle of the calendar can also make it feel as though time is rocketing by. Each year, the two of us look out on the first day of school and marvel at how quickly the college campus fills up with students following a blissful summer of quiet. (Don't misunderstand: We love our students, but there is something idyllic about a nearly empty college campus.)

On that first day of class, when the campus fills with bustling crowds, we also feel mentally closer to all of those other first days of class in all our years of teaching. Decades of teaching experience are compressed into a single day, and it feels as though the years have flown by.

All of us experience similar cycles in life. It might not be related to the first day of school, but rather a holiday or a birthday or a wedding anniversary. On those occasions, the current experiences serve as cues that prompt memories of similar events in the past. The more detailed your memories, the nearer they seem in space and time, so it's possible that the years separating the present from the past don't seem so long.

Lives filled with new experiences are less likely to seem like a blur in the rearview mirror of our memories. So in the midst of our life experience, it may be possible to create the kinds of memories that lead to more satisfying perceptions of the past—perceptions that are bolstered by memories of new experiences. Keep trying new things. Find new hobbies. Listen to new music. Make new friends. Read new books. Travel to new places.

Every time you add something new to your life, you create the conditions for future remembering—thoughts of years that were full and meaningful. But even though you can add richness to your days in ways that make your life feel full, even when you look back on it,

Time flies WHEN YOU'RE GETTING OLD.

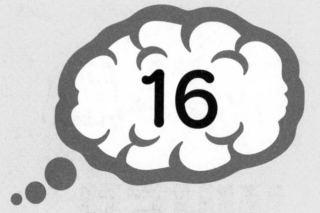

Why is forgiveness
so powerful?

UNLESS YOU HIDE AWAY IN A CAVE FOR YOUR ENTIRE LIFE, IT IS INEVItable that other people will wrong you in some way. Some of these personal offenses may be minor. Friends of yours who are getting together to see a movie that you're interested in seeing may forget to invite you. It may sting a little, not to be included in the group outing, but the world won't end because of it.

Other transgressions are more serious. The singer-songwriter Leonard Cohen spent years living in a Buddhist monastery. While he was away, his manager stole millions of dollars of his money. As a result, the singer had to go back on tour in his seventies in order to earn a living.

It would seem strange to carry around a grudge for being left out of a trip to the movies, but it would certainly seem reasonable for someone whose life savings had been stolen by a trusted advisor to remain angry at the thief.

At one time or another in all relationships, one person does something (or doesn't do something) that causes hurt in another person. Sometimes these actions or omissions are not intended to cause bad feelings. At other times that's precisely what they're intended to do. After events like this, relationships often change.

Imagine two people—Art and Bob, for example—who get together periodically to do a radio show. Art arrives at the studio one morning with breakfast tacos and coffee for the show's producer Rebecca and their engineer David, but with nothing for Bob. At this point, Bob is well within his rights to feel insulted. He might even look for subtle ways to express his annoyance.

(By the way, this might seem like a minor transgression—particularly for non-Texans who have never experienced the joy of the breakfast taco—but breakfast tacos are a truly heavenly experience involving breakfast foods like eggs, avocados, beans, et cetera, all wrapped in a small tortilla. Leaving someone out of the breakfast taco experience is a major faux pas. And Art knows that. And Bob knows that Art knows that.)

At this point, there is a slight tear in Art and Bob's relationship (and a tear in Bob's eye). We might expect Art to apologize for overlooking Bob's dire need for breakfast tacos and caffeine. An apology would signal that Art recognizes he has done something wrong. It is also a signal that he is less likely to do something like that again in the future, because he is aware that his actions were hurtful. If Art has generally been reliable, though, Bob might be more inclined to take Art's apology seriously. On the other hand, if Art has a history of slighting Bob, then Bob may not want to listen to Art apologize (yet again . . .).

When Bob decides to forgive Art (which he always does), his doing so sends a signal to Art that the past infraction is no longer going to be a primary factor that influences current interactions. With his forgiveness, Bob is signaling that he is willing to return (more or less) to the state of the relationship before Art was so thoughtless. This signal obviously helps Art because now he can begin to focus on the present when interacting with Bob, rather than having to be reminded of his mistake every time they encounter each other.

Studies demonstrate that people have a hard time forgetting the details of other people's transgressions when they have not forgiven them. When people forgive the transgressor, though, the forgiving actually leads to their *forgetting* the details of what happened in the past. So from the time that Bob forgives Art, the specific details of that horrible day when Art didn't bring breakfast tacos begin to fade. That allows them to get along well again (until Art forgets the guacamole).

ONE REASON THAT BOB MAY BE WILLING TO FORGIVE (BEYOND HIS KIND AND generous nature) is that he is older than Art. Research suggests that older people are generally more forgiving than younger ones, and there are a couple of reasons why this is true. One is context: When you are young, it's difficult to know which of the various insults you've suffered are really going to matter for the rest of your life. As you get older, you begin to realize that many of the bad things you've suffered are not really that bad. While you may have perceived something as a major offense at an earlier stage of your life, that same offense often seems to have little lasting impact on the quality of your life when you observe the situation from a more seasoned perspective. So it tends to get easier to forgive and forget.

Another reason older adults are more likely to forgive than younger adults is that there is drift in people's personalities as they get older. One core personality dimension (another one of the Big Five) is *agreeableness*, which is the extent to which people really want to get along with others. A second key dimension (yet another of the Big Five) is *neuroticism*, which is the amount of anxiety, stress, and emotional energy people experience regularly. As people get older, they tend to become more agreeable, making it more likely that they will want to get along with others. They also become less neurotic, and therefore tend to be less stressed and anxious about life. These two personality changes also make it easier for people to forgive others who have done them wrong.

Of course, not every bad act deserves forgiveness. Leonard Cohen testified at the trial of his former manager and talked about how he was harassed in addition to being stolen from. He pursued justice, and ultimately his manager was jailed for her crime. He was sensible enough not to simply forget the incident and move on without first seeing his manager punished.

But most wrongs that are perpetrated in personal relationships are not resolved in courts of law, and we all benefit from reaching some sort of resolution. In personal relationships, such resolutions begin with acts of forgiveness. Skillful therapists often work with their clients to help them forgive people who have done (sometimes horrible) things to them in order to help them move forward with their lives and to avoid being trapped in an endless cycle of rumination about painful actions that can't be undone. In this way, forgiveness provides a way to move past the negative emotions that come with the resentment of past wrongs, encouraging people to regain a sense of optimism and control over their lives.

Forgive and forget (LITERALLY).

Is our
thinking
ever coherent?

Aᴸᴸ ᴏꜰ ᴜꜱ ᴀʀᴇ ᴘᴇʀꜰᴇᴄᴛʟʏ ᴄᴏɴᴛᴇɴᴛ ᴛᴏ ʜᴏʟᴅ ʙᴇʟɪᴇꜰꜱ ᴛʜᴀᴛ (ᴀᴛ ʟᴇᴀꜱᴛ on the surface) don't seem to be consistent with one another, and incoherence is a common feature of human thinking. Yet there are situations in which our beliefs are more coherent.

A few years ago, it was time for Bob to buy a new car. He wanted something that would be convenient to park around the campus and would get good gas mileage. Bob is not what you would call a "car guy." He likes to drive, but he doesn't need a high-performance sports car (as Art does . . .).

He looked at several different cars, and they all had their strengths and weaknesses. The Mini Cooper® was cute and handled well, but it was pricey. The Fiat® was small, easy to drive and park, and less expensive, but was also just newly introduced in the United States, so it wasn't clear how reliable it was. The Smart® car was tiny, got great gas mileage, and was easy to park, but it was also constantly under threat from the gigantic SUVs that roam the streets of Austin.

Slowly, Bob started thinking that the Smart car might be the car for him. As he began to reach that conclusion, some interesting things began to happen. He became more focused on the convenience of a small car and the importance of energy efficiency as well as price. Over time, he got less worried about the size of the other cars on the road. That is, as he began to develop a preference for the Smart car, he also changed his beliefs about what was most important about his car purchase in general. These shifts in his belief system supported his growing decision to buy a Smart car.

Bob's experience is typical of the spreading coherence that is common in many aspects of thinking. How can there be spreading coherence if we are happy to have lots of beliefs that are inconsistent?

To understand this, we need to think about a basic distinction in types of memory. There are lots of things that you know. For example, you might know that the president of the United States when the 9/11 attacks occurred was George W. Bush. You might know that Roger Bannister was the first person ever to run a mile in under four minutes. You might also know that water has the chemical symbol H_2O.

Although you may know these things, you probably weren't thinking about any of them a few minutes ago. Instead, they resided in *long-term memory*. Long-term memory comprises all of the stuff you know: facts you have heard, stories you have been told, visual memories of past life events, names of people you have met. The amazing thing about long-term memory is that you may remember some things for many, many years. Indeed, people in their eighties can often recall details from childhood.

THE MEMORIES THAT YOU HAVE IN LONG-TERM MEMORY NEED NOT BE CONSIS-tent with one another. You may never recognize that *Look before you leap* and *He who hesitates is lost* seem to offer opposing prescriptions, even if you know both proverbs. They reside happily together in long-term memory without ever causing any concern about the fact that their meanings conflict.

After I mentioned George W. Bush, Roger Bannister, and H_2O, this knowledge was drawn from long-term memory into *working memory*. As we mentioned earlier, working memory is the information that is relevant for whatever you are thinking about right now. While reading this book, lots of psychology concepts (like memory, attention, habits, and personality) are likely to enter working memory. More distant information like romaine lettuce, Hula-Hoops®, and Rice Krispies® are not (at least until we mention them).

Working memory is limited. There is a relatively small amount of what you know that actually influences what you are thinking about at any given moment. When pieces of information enter working memory together, though, there is pressure on you to evaluate them in a way that is consistent. Bob's beliefs about the cars he was considering became more coherent with his evolving preferences over time because he was thinking about different cars and their features at the same time. In other words, information about the cars was active in working memory rather than just being stored away in long-term memory.

As a result, knowledge that was consistent with Bob's develop-ing preference for a Smart car increased in its goodness and impor-tance. Information inconsistent with his preference for a Smart car decreased in its goodness and importance. After a while, Bob was focused mostly on those aspects of the Smart car that convinced him Smart cars are good to own.

Similarly, while you would have been content to believe both proverbs I mentioned earlier, if you hold both of them in working memory at the same time, you will recognize the contradiction between them, which will prompt you to start resolving that contradiction.

THE FORCES THAT CREATE COHERENCE AMONG PIECES OF INFORMATION ACtive in working memory affect all kinds of opinions—not just preferences related to choices. Classic examples of *cognitive dissonance* work the same way. The idea behind Leon Festinger's theory of cognitive dissonance is that when you contemplate two contradictory beliefs at the same time, doing so makes you uncomfortable, so you find a way to resolve the contradiction. You may not resolve it consciously, but psychological mechanisms work to make your beliefs more coherent.

For example, Art has a very high opinion of himself (as if that was not obvious by now). Yet he has not won any significant awards in his career. If Art really were such a big shot, then you would think he would have won *something*. This thought could potentially be very distressing for Art. After all, Bob has won a number of important teaching awards.

The way that Art deals with this threat to his self-esteem is to decide that awards are not really that important after all. He is certainly happy for his friends and colleagues (like Bob) who win awards, but he recognizes that these awards don't really mean much (at least until he wins one). In this way, Art can know all of the relevant facts about awards (they exist, his colleagues have won many, he has not) and still manage to maintain his self-esteem.

The consistency of the beliefs that enter working memory involves facts, but it can also involve feelings and emotions. Think about examples like Bob's car purchase. Suppose he had seriously been considering purchasing a Fiat, and then he found out that a hated administrator just bought one. That might make him feel bad about Fiats. This negative emotion would also affect his beliefs about the features of Fiats, so he would start to believe that the good features of Fiats are less important and that factors like the lack of information about reliability are big problems.

IN THE END, THEN, WE ONLY LOOK FOR CONSISTENCY AMONG FACTS THAT ARE currently in working memory at more or less the same time. This mechanism of consistency leads to the *change of standard effect*. We use lots of descriptions for people that imply a comparison between a given person and others. Suppose, for example, that you have a friend in college you think is particularly insightful. Her arguments always seem very clear, and she brings up points you have not thought of before. Years later, whenever you think of your friend, you think of her as being insightful.

Later in life, after not seeing her for many years, you meet up with her again. To your surprise, you find that your conversations with her are enjoyable, but she doesn't seem to have any special way of looking at new situations. What happened? Consider the possibility that when you originally labeled her as insightful, you were comparing her with yourself and with other people you spent time with in college. Compared with that group, she was insightful, but as your experience in the world grew and you interacted with more people, your definition of what it means to be insightful changed.

The "insightful" label you assigned your friend lived on for so many years, though, because you had no opportunities to update your beliefs about her. You hadn't been thinking about her since college, so there was no way to compare her level of insight with that of the people you interact with more frequently now. That is, the label you gave her resided in long-term memory and was not actively compared with your beliefs about other people until you met her again.

All of this suggests that there is a premium on consistency of thoughts and beliefs *when you are thinking about things in the moment*, but not when information is stored away for the long term. In other words:

COHERENCE =

current +

CONSCIOUS +

COMPARISON

Are our beliefs consistent with one another?

A T ONE LEVEL, MOST PEOPLE'S THINKING CAN SEEM RATHER JUMBLED and incoherent. For example, there are all kinds of proverbs that we think have merit, even though many of them convey contradictory messages. For every proverb (say, *Look before you leap*), there is another that expresses the opposite sentiment (*He who hesitates is lost*). In different circumstances, we are quite willing to accept that both of these proverbs are true.

In fact, belief systems often have significant contradictions built right into them. And it isn't just because we have learned lots of proverbs.

For example, in the United States in the early twenty-first century, it is common for people on the political "right" to be opposed to abortion and to allowing terminally ill patients to commit physician-assisted suicide, and yet many still support the death penalty for convicted murderers as well as the use of deadly military force with perceived enemies. They defend their beliefs about abortion and physician-assisted suicide by appealing to the sanctity of life, but carve out exceptions for punishment and national security. On the political "left," it is common for people to favor individual freedom when it comes to smoking marijuana and marriage for gays and lesbians, but to favor restrictions on hate speech and gun ownership. Thus, the belief that individual freedom is a crucial aspect of American society extends only to a subset of behaviors.

To be clear, we are not evaluating the merits of these beliefs. But it is interesting to observe that individuals can claim to have core principles (like the sanctity of life or the importance of individual freedom) that guide their beliefs, while at the same time holding opinions that are inconsistent with those principles and about which they are often unaware.

Part of this problem is pragmatic, of course. A core principle that you hold and do not want to have violated is called a *protected value*. When people have a protected value, they don't even like to consider violating it. Watching other people violate protected values can cause feelings of anger and even outrage. When individuals contemplate violating their own protected values, they feel guilt and shame.

For example, Art grew up in New Jersey in a family of New York Mets fans. (Bob was lucky enough to avoid strong sports allegiances as a child.) Mets fans are the kinds of people whose two favorite teams are the Mets and whoever is playing the New York Yankees. As a result, hating the Yankees is a protected value. Indeed, as a kid, just talking to a Yankees fan was nearly enough to make Art's blood boil. (Happily, he has mellowed as an adult and is proud to say that he has several friends who are Yankees fans, misguided though they may be.) In addition, on the one occasion that Art went to Yankee Stadium for a game, he felt guilty for a long time. Actually, he is still a little ashamed of it.

The thing is, once you have more than one protected value,

those values are very likely to come into conflict at some point. People who oppose abortion and physician-assisted suicide, but who favor the death penalty for murderers and deadly military force for regimes perceived as threats to American lives and values, are experiencing this kind of conflict. They have two deeply held values—the sanctity of life and the prime importance of security—and different circumstances require making a choice between the two.

What is fascinating, though, is that such choices are rarely explicit. Indeed, most people are not aware of the inconsistency of beliefs like this until it is pointed out to them. (We will have more to say about protected values toward the end of the book.)

And that brings us to the first point about coherence: Your brain doesn't care about it.

In other words, if you learn some new fact that turns out to be inconsistent with something else you know, there are no automatic mechanisms in your brain that point out the inconsistency and force you to resolve it. Instead, you simply end up with two different beliefs that are not consistent.

How can that be?

WELL, ALMOST ANY BROAD STATEMENT YOU CAN MAKE ABOUT HUMAN BEHAVIOR is true only in certain circumstances. The trick to understanding behavior is to know the circumstances in which behaviors are going to happen.

The same thing is true with beliefs. When someone says, "I believe that human life is sacred," or "I believe in individual freedom," that statement includes an unstated disclaimer like "all else being equal." But there are nearly always circumstances that lead to the violation of any broad belief or value statement.

It would be too much work for the brain to have to enumerate all of the exceptions to the rules you believe in, so instead the brain does something easier: It associates beliefs with specific situations and makes it easier to retrieve those beliefs in the situations with which they are associated.

Suppose Bob travels to a national park. There are signs all over the park warning people to beware of bears, so Bob learns that he should not go near a bear and that he should be afraid of them. But then Bob goes to the zoo. There is a bear at the zoo as well, but Bob need not be afraid of the bears at the zoo because there are moats and fences to protect him.

Theoretically, Bob could learn the rule "Be scared of bears," and then learn all kinds of exceptions to that rule. Or he could learn the rule as well as the context in which that rule was learned. That makes it easier to recall the information in that context again in the future.

Most of the time, Bob is unlikely to encounter a bear, so he won't think much about bears at all. In national parks (and things that look like national parks, such as forests), Bob might encounter a bear, though, so it is valuable to be able to retrieve information about the dangers of bears in that setting. When Bob thinks about bears in zoos, on the other hand, he is not scared of them because the lesson he learned in that setting is that the zoo's reinforcements protect people from unwanted encounters with bears.

Because this system works pretty well, most of the time you do not need to think about the fact that your beliefs may be contradictory. But calling to mind your contradictory beliefs leads you to notice that they are not consistent. (And there seems to be an endless reservoir of people who delight in pointing out your inconsistencies to you, particularly on the Internet.) In those situations, you have two options.

One is to follow the "it depends" strategy. In this case, you make a mental note that your beliefs are not really contradictory. Instead, one belief holds in one set of circumstances, and the opposite holds in other circumstances.

Sometimes, though, you resolve the conflict between beliefs by choosing one over the other. This strategy is the one we use in science. When we do a scientific study, there are often competing theories that attempt to explain some aspect of the world. When two theories are inconsistent with each other, we use data to decide which one we ought to believe. Relying on the collection and analysis of data to determine whether theories are wrong is a protected value in science. The whole process forces conflicting ideas into stark juxtaposition in an effort to resolve conflicts.

Individuals are less often forced into such quandaries on a day-to-day basis. One belief can happily coexist with other conflicting beliefs until someone or something highlights the contradiction. The resulting dissonance in some cases may lead to a careful re-examination of values, or it may lead to an expedient rationalization and a quick change of topic. The interesting thing in all of this is how effortlessly we can hold disparate beliefs, even when challenged.

Walt Whitman perhaps offered the best rejoinder: "Do I contradict myself? Very well then, I contradict myself (I am large, I contain multitudes)." Good one.

The most important words in the field of psychology are

IT DEPENDS.

Why is it hard to learn a new language?

MOST PEOPLE IN THE UNITED STATES ARE REMARKABLY MONOLINGUAL. It's hard to get a good estimate of the number of adults who feel proficient enough in a language other than English to have a conversation with a non-English-speaker—estimates range from 10 percent to about 30 percent—but it seems safe to say that only about one in four adults in the US speaks a second language well.

It is not for lack of trying, though. Most students in the United States are enrolled in second-language classes in middle school or high school. Yet even though many children study a second language for several years, most don't really learn to speak it.

Why is that?

To put this problem in perspective, it is worth looking at what happens with children who are exposed to multiple languages in early childhood. Children who live in homes where two languages are spoken may start speaking somewhat later than children who grow up in single-language homes, but they learn both languages effortlessly and grow up to be fluent speakers of both languages, provided they continue to use them. Not all bilingual children become bi-*literate*, however; that is, they may not learn to read and write in both languages, even if they are fluent speakers.

Other children grow up in households where only one language is spoken in the home, but they are introduced to a second language outside the home at a very young age. This is most common in countries with a regional language that is generally spoken among the people and an official language of the country that is taught in schools. In cases like this, children may not be exposed to the second language until the age of four or five, but because they have contexts in which they need to use each language, these children also become fluent bilingual speakers.

A surprising observation, though, is that the closer children get to puberty before they are exposed to a new language, the harder it is for them to achieve native fluency. There are several factors that lead to this difficulty, the first of which involves statistics.

A lot of language learning involves noticing the statistical distribution of speech sounds. All languages comprise a number of different speech sounds, called *phonemes*, and human brains, even in infancy, begin to calculate how likely it is that certain phonemes will be followed by other phonemes. Anyone who's watched a foreign-language film knows that sentences in an unfamiliar language sound like long uninterrupted strings of phonemes. It's nearly impossible to tell where one word ends and the next one begins, but the remarkable calculations in the infant brain—which occur quite automatically—help babies learn about the basic structure of language before they even understand what's being said.

Grammar involves the ordering of words, the endings that

make nouns and verbs "agree" in speech, and all of the factors that allow you to string words together into sentences to create new sentence meanings. Statistics comes into play here as well.

Most native speakers reliably follow the rules of language usage, but few native speakers can actually state what the rules are. Even though you are able to generate all kinds of new sentences, you probably find it difficult to state all the rules that allow you to put one word after another. Doing so is difficult because you first learned the rules for generating sentences *implicitly* (without conscious awareness), just by hearing and using language over time.

Children use language, of course, but they don't spend much time thinking about language explicitly. They're ill equipped to do so. The frontal lobes of the brain, which are involved in the careful analyses of conceptual material, are not well developed in young children, so instead of approaching language learning through analyses of what's going on, children just focus on trying to communicate with other people.

Those regions of the frontal lobes become increasingly more developed as children approach puberty (although they don't finish developing until you reach your early twenties). So if you start learning a language when you are in your teens or later, you tend to approach the language through analysis of what you're trying to do. Rather than just letting the language wash over you and allowing your brain to covertly calculate all kinds of statistics from what you are hearing, you focus on learning rules about which kinds of words follow other words in sentences.

Rules of language might be great for answering questions on a high school language exam, but they are not so good for actually communicating quickly and effectively. Paradoxically, the rapidly developing critical thinking skills of teenagers actually make it *harder* for them to learn a language than it would be if they were not so adept at analyzing what they are doing.

A SECOND FACTOR THAT MAKES IT HARDER FOR KIDS TO LEARN LANGUAGE once they reach puberty is embarrassment.

No, really—embarrassment.

One of the most important requirements for learning a language is to actually use it. Engaging in conversations with native speakers allows you to hear the way people speak and to experience a native accent. In order to participate in these early, bumbling conversations, you need to learn some essential words and practice using them in sentences. But if you have spent any time with middle school and high school students (not to mention adult colleagues and friends), you know that they are very concerned with their social status. Embarrassing themselves in front of their peers is not something they relish doing. However, trying to say something in a new language is a surefire way to cause at least a little embarrassment. Chances are you will use words incorrectly, mispronounce them, and make grammatical errors. You might even do all of that in the same sentence!

And the surest way to avoid mistakes, especially in front of other people, is not to say anything at all.

All of this leads to the inescapable conclusion that, here in the United States, we start teaching languages to kids at the age when they least want to use a new language to communicate. Because adolescents generally avoid making mistakes in front of their classmates, they tend not to use the language much, and therefore learn much less about how to actually speak a second language.

A third factor that makes learning a new language difficult has to do with the sounds (phonemes) that different languages use. In English, for example, we make a distinction between the sound that goes with the letter r and the sound that goes with the letter l. Japanese does not use this distinction. The sounds that we distinguish as r's and l's in English are part of a single phonemic category in Japanese, so when a native Japanese speaker hears the sounds of r and l in English, they seem like different versions of the same sound.

The resulting mispronunciations are not merely a result of the inability to make lips and tongues do the tricks necessary to *produce* the sounds, but also a result of not being able to *perceive* the differences between them. Many of the variations among phonemes in a given language are imperceptible even to native speakers. For example, if you say the word *pit* and the word *spit*, the *p* sound you make is actually different. Never noticed that before, did you? There is a puff of air that accompanies the *p* in *pit*, but not in *spit*. Stick your hand in front of your face as you say each word. You can actually feel it. In English, that puff of air has no meaning, and so you hear the sounds as variants of the *p* sound. In languages like Korean, though, these sounds are used to distinguish between words, so Koreans are sensitive to this difference.

Other sounds that are common in other languages aren't used in English at all. For example, the German word *Ich* ends with a phoneme that may sound to an English speaker like someone clearing her throat. Hebrew and Arabic use a similar sound, but English does not. The challenges for English speakers in these instances are primarily problems of production. Learning to produce speech sounds you've never made before is a heavy lift.

Quite remarkably, babies are born with the ability to distinguish among the speech sounds that are used to create words in every human language. The human auditory system comes equipped to do that (at no extra charge). Over time, though, you lose the ability to make distinctions like the ones we just described if they are not meaningful in your native language, and if you then try to learn a language in which those sounds *are* meaningful, you will have a hard time speaking in a way that native speakers understand clearly.

FINALLY, THERE ARE ASPECTS OF LANGUAGE THAT TRULY DEFY RULES. WHAT we mean is that some aspects of usage are not based on rules at all; you just have to memorize how they work.

For example, you probably haven't spent much time thinking about *prepositions*—words like *in*, *on*, and *at* that specify location or action in relation to another entity. You also probably think you understand exactly how prepositions work in English, and if you're like most people, you find nothing about them particularly strange. And yet they are actually quite strange . . .

Consider the preposition *on*. You use this word to talk about objects on top of surfaces. But sometimes those objects are supported by the surface (the apple is *on* the table), while at other times those objects are simply in contact with the surface (the picture is *on* the wall). To make matters more confusing, there are times when something is supported by a surface, but a different preposition is needed (the apples are *in* the bowl).

To make matters yet more confusing, the prepositions of every language are a little different. While English uses the same preposition for an object on a table and one hanging on a wall, Dutch uses different prepositions for these two relationships. So even after you manage to master when you are supposed to use each preposition in your native language, there's no reason to expect that you will figure out how to use them in a new language. In fact, adults who learn a new language make more mistakes with prepositions than with just about any other aspect of speech.

Like grammar in general, the best way to learn the prepositions of a language is to hear them, use them, and allow your brain to recognize which ones are appropriate in different circumstances by taking into account both the meaning and the statistics of when they are used. This kind of implicit learning requires a lot of exposure to the language, which is hard to get unless you are willing to really become a part of a community that uses it.

All of this suggests that language learning is both easy and hard to do. It is actually quite easy for kids and gets more and more difficult with increasing age. By the time you reach puberty, you are unlikely to ever speak a new language like a native. That doesn't mean that you shouldn't bother to learn a new language as an adult. In fact, learning a new language, like many intellectual challenges, is a tremendously effective way to engage your brain and accrue the many benefits we've mentioned elsewhere in the book. But the fact remains: Learning a new language as an adult is different, and more difficult, than learning a language as a child.

Rather than trying to analyze your way into perfectly uttered sentences, immerse yourself in the sounds of the language and avail yourself of opportunities to speak it, even (especially) when you're not quite sure what you're doing.

In sum, if you want to learn to speak a language like a native speaker:

DON'T THINK; LISTEN AND SPEAK.

Is our right brain different from our left brain?

I F YOU LOOK AT A BRAIN, IT SEEMS PRETTY SYMMETRICAL. THE RIGHT AND left sides have the same general shape. The ridges on each side of the brain fall in about the same place. Yet you have probably heard people talk about being "left-brained" or "right-brained." In these conversations, the left brain is associated with being logical and adept at using language. The right brain is associated with being creative, artistic, or musical.

What is going on here?

The origins of this distinction come from studies that were done by Roger Sperry (a Nobel Prize winner) and his student Michael Gazzaniga (who went on to become a prominent cognitive neuroscientist). Sperry's work suggested that one way to treat severe epilepsy was to cut the thick band of fibers, called the *corpus callosum*, that separates the two halves (or hemispheres) of the brain. Sperry and Gazzaniga proposed that the epileptic seizure would be restricted to the half of the brain from which the seizure emanated, leaving the other hemisphere unaffected.

And they were correct: The surgery was successful at limiting the strength of seizures—but it had other important consequences for the patients on whom it was performed. Cutting the corpus callosum meant that almost no signals could travel from one side of the brain to the other, which turned out to be a big problem, because not all of the information about the perceptual world is available to the whole brain immediately. In vision, objects that are on the left side of your visual field are initially processed by the right side of your brain, while objects that are presented on your right are initially processed by the left side of your brain.

We should clarify that when we say "on the left side," we mean to the left side of your body—not your left eye. Light from the left side of your visual field goes into *both* the left and right eyes and strikes the right side of the retina (the carpet of light-sensitive cells at the back of your eyes)—in each eye. Activity from the right side of both retinas goes first to the right hemisphere of the brain. Similarly, light bouncing off objects to your right strikes the left side of both retinas, and activity from the left side of the retinas goes first to the left hemisphere of your brain.

In a normal brain, signals from the right and left hemispheres are exchanged quickly across the corpus callosum. But for patients whose corpora collosa were severed, the signals could not cross from one side of the brain to the other, so visual information about what was seen remained in the hemisphere where the signal was first received.

Sperry and Gazzaniga conducted a number of experiments on the patients who had undergone the procedure. While facing forward, patients were shown printed words that were flashed in either the left or right side of their visual fields. When a word appeared on the right side (so that it was processed by the left side of the brain), patients could report the word they saw. But when the word appeared on the left side (so that it was processed by the right side of the brain), patients could not report the word.

These studies were the first indication that the hemispheres of the brain were in some ways specialized for different tasks. Indeed, in most right-handed individuals, the language centers are primarily in the left hemisphere of the brain. This finding related to earlier studies of neurological patients showing that strokes on the left side of the brain were far more likely to disrupt language than were strokes on the right side.

Notice, however, that we said *right-handed* people. Left-handers (like Art) are a little strange (at least in terms of their brain organization). Some left-handed people also have language centers mostly in the left hemisphere. Other left-handed people are opposite, with language dominant in the right hemisphere. Still other left-handers have language centers on both sides of the brain. That is why, if you look at many studies that examine language and the brain, they include only right-handed people.

It is important to point out that, even in these initial studies, there was evidence that the right hemisphere of the brain is also involved in some ways in the processing of language. For example, if patients saw a word flashed on the left side (so that it went to the right hemisphere), they could reach into a box of objects with their left hand and pull out an object whose shape was described by the word. If they saw the word *star*, they could pull a star out of the box, even though they couldn't say the word itself. But in order to select the correct object, they had to use their left hand, because the left hand is controlled by the right hemisphere of the brain.

SINCE THESE INITIAL EXPERIMENTS, THERE HAVE BEEN MANY MORE STUDIES OF hemispheric differences in the brain that have revealed a number of interesting features of perception and thinking. When many people listen to music or think about music, for example, they show dominant activity in the right hemisphere of the brain. Findings like this are fascinating, and they have given us a lot of information about how the brain is organized.

They have also captured the public imagination. As the growing understanding of hemispheric specialization became popularized, many specific, carefully articulated reports of research were overly generalized and often misapplied. There are still many who argue that schools have been focusing too much on children's "left brains," emphasizing mastery of language, science, and math, and not enough on children's "right brains," which was taken to mean more holistic and artistic skills. Many people even describe themselves as mostly "left-brained" or "right-brained" to mean analytical or holistic/creative.

In fact, in the normal brain, information is shared between hemispheres all the time. Even though there is some specialization of the functions in each hemisphere, normal thinking involves the coordinated activity of the two hemispheres. In simple terms, no one thinks only with the left or right side of the brain.

It's not even clear that music is processed only on the right side of the brain (in right-handed people). A great deal of evidence illustrates that as people gain expertise in things like music (with a right-hemisphere dominance), the left side of the brain becomes increasingly involved. Arguments in favor of children's music study in school (which is great for a number of reasons) because it "educates the right side of the brain" misunderstand this aspect of brain function.

The organization of the brain is highly malleable, and learning experiences alter the way the brain works. Even ascribing language processing to the left hemisphere may in part be a result of expertise. All adults are language experts, and the "language centers" in the left hemisphere probably reflect a level of expertise more than the type of information being processed (like music versus language).

There are even more subtle explanations for the differences in the hemispheres. Some theories suggest the right and left hemispheres differ in whether they process fine details or broad summaries. For example, if you squint when you look at the world, you see only the broad summary of what's in view, but if you keep your eyes wide open, you also see the fine details. It may be that language centers tend to form on the left side of the brain because learning and using language requires paying attention to fine details of speech sounds and word order.

All this suggests that the popular uses of the terms *left-brained* and *right-brained* are not quite right. People who are more scientifically inclined or who use more math are not thinking mostly with the left side of their brains, and people who are artistic, musical, intuitive, or theatrical are not thinking mostly with the right side of their brains. In fact, if the theory that the left hemisphere focuses on more finely detailed information than the right hemisphere is correct, then both sides of the brain are engaged in nearly every task you perform.

Healthy thinking requires taking many approaches to all of the problems we solve in life. It is unfortunate that there has been a move to cut the arts and humanities from a number of school curricula, because the role of disconfirmation that emerges from science (which we discussed in an earlier chapter) needs to be complemented by an understanding of how broad themes weave through human interactions (as you might learn from history or literature).

For example, no matter how good a given technology may be,

people won't buy it unless they can envision how to use it and unless the design of the product fits with the way people think. That is why there were many unsuccessful MP3 players before the iPod®: It took the genius of the Apple® design team to recognize how to make the technology easy for the average person to use and simultaneously pleasing to the eye.

Indeed, the most successful scientists and mathematicians have a finely honed sense of what is beautiful. We both attend conferences with fellow scientists all the time, and we routinely hear scientists describe theories as "elegant" or "clunky" and data that are "beautiful" or "ugly." These aren't metaphorical descriptions. As scientists gain expertise, theories and patterns of data create feelings that are pleasing or not in the same way that listening to music can make someone feel transported or punished.

There are very good reasons to include a mix of science, humanities, and the arts in children's education. Each of these disciplines teaches different approaches to thinking and provides valuable skills.

To answer the question we posed at the beginning of this chapter: Yes, there are some differences between the right brain and the left brain, but it's a moot point.

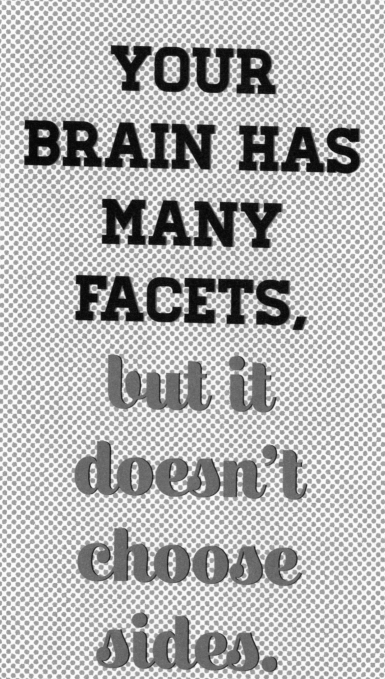

YOUR BRAIN HAS MANY FACETS, but it doesn't choose sides.

21

How do we overcome writer's block?

O KAY. WE THINK WE'VE GOT SOMETHING NOW. WRITER'S BLOCK IS that most desperate condition experienced by authors who become so stuck in their efforts to write that they are unable to compose anything they're satisfied with. Sometimes they simply sit and stare without writing anything at all. Or they get mired in a futile cycle of writing a little, groaning, and deleting that seems to go nowhere good.

What causes writer's block? How can people who clearly know how to write suddenly have nothing that seems worth writing?

It certainly isn't a sudden loss of the ability to use language. Writers who can't seem to make any progress on the page continue to talk with friends and family, creating sentences on the fly that have never been spoken before in the history of human language— so it's not as if they have lost the ability to create sentences. But they just can't seem to move forward with the project they are working on. What's up with that?

In fact, writer's block comes from *anxiety*. Every writer wants to generate good ideas and express them effectively. Writers get into a funk when they feel they can't come up with brilliant new ideas or they can't write them with enough clarity, beauty, or wit.

It is probably important to say up front that Art has no real first-person understanding of writer's block. He tends to fill up any blank page with words at a frightening clip. Many of those words make no sense at first, but that has never stopped him. Art's recklessness with the written word is actually a helpful example for understanding how to overcome writer's block. Art could not care less how bad the sentences or ideas are that emerge when he starts to write, but he *does* care a lot that things get written at all.

Bob likes to point out that most writers compare what they are writing *as they are writing* with the finished products of great writers they have read in the past or even with past examples of their own best efforts. A fiction author might wonder why her sentences don't read like Hemingway or Updike (or even Stephen King or James Patterson). Again, the problem here is the evaluation of early drafts as if they're finished products that have gone through multiple edits by multiple editors.

Of course, most readers see only examples of what gets published. What they *don't* see are all the drafts that went into creating these finished works: the horribly constructed sentences that needed to be taken apart and put back together again; the tortured logic that scuttled an argument; the use of the same word repeatedly in sentence after sentence.

Few writers churn out highly polished work on their first drafts the way Mozart reportedly spun out symphonies. But most successful writers learn a valuable lesson that is also supported by a lot of psychological research: The people who have the *best* ideas are the ones who have the *most* ideas. That is, there isn't really a great way to generate *only good* new ideas. Instead, you just have to crank out a lot of stuff and then learn to spot the good ideas when they present themselves.

The trick to doing anything creative (like writing) is to be unafraid of bad and mediocre ideas. Just get stuff out there. It doesn't matter if it isn't perfect. You might end up throwing out everything you did on a particular day. When Art wrote his book *Smart Thinking*, he had two completely different finished drafts of one chapter that he trashed before he finally wrote a third one that he (and his editor) liked enough to keep. If he had waited until he had a perfect draft, he might have given up altogether.

Actually, this story illustrates another important point, which is that you also shouldn't be afraid to show other people your ideas, even when you have no confidence in them (the ideas, not the people). Sometimes a friend or editor will hate what you wrote and you'll have to redo it, but that's no reason to keep your drafts to yourself. Quite the contrary: It is through the process of responding to criticism—their own and others'—that writers ultimately develop their best work.

One reason people are afraid to show their ideas to other people is that they suffer from a variant of *imposter syndrome*. That is, they feel like they are the only one among their peers who struggles to do things competently. They worry that if other people knew how much effort it took them to do something that is merely acceptable, then those people would recognize them as the frauds they really are.

The thing is, almost all people curate their public face. Look at people's Facebook pages. Every family is beautiful and smiling. Every kid is doing well in school. People don't generally post the mistakes they make or the problems they are experiencing. What you see of other people is what they are choosing to show, and they don't usually decide to broadcast all their flaws, errors, and failed ideas.

A big part of being successfully creative is overcoming the sense that you don't really deserve to be thought of as a creative person. All writers have constructed horrendous sentences and entire passages that don't make sense. They have written boring stories that are completely derived from something written by someone else. They have written incoherent statements from time to time.

Other people are probably not going to think badly of you if you show them something that isn't perfect. They might even be a little relieved that you—like them—are human. They might even offer some advice to help you make it better.

When worrying about what other people are going to think of your work leads you to edit all your ideas before you even give them a chance, you miss opportunities to consider possibilities and refine your thinking. Sometimes a bad or mediocre idea is just a stepping-stone to a better one.

The only way to generate an idea is to draw it out of memory somehow. Information comes out of memory in response to whatever you are thinking about at the moment. For example, think about a class you attended at some point in your life. That wasn't so hard to do, was it? We asked you to think about a class, and you did.

If you want to change what you are thinking about, you need to ask your memory a different question. One way to do that is to let those seemingly bad ideas out first. You just might find that the initial, perhaps bad idea makes you think of something else that is a little better. Over time, the terrible idea you started with becomes a good idea—perhaps even a brilliant one.

In order to keep writer's block at bay, writing needs to become a routine in which you sit down, get things out, and then edit them after that. The prolific author Stephen King seems to come out with a new book about every twenty minutes (oh look, there's another one). He has also written a lot about his own writing process. His big secret? He writes every day. Each morning, he gets up and writes for a few hours. He suggests that writers need to develop habits around writing just like everyone needs to create a routine around going to sleep at night.

One reason to write every day is that writing is a skill. Bob spends a lot of time teaching musicians and helping musicians learn. Eventually, all musicians come to recognize that they have to keep practicing in order to get better at what they do. They also have to get lots of feedback from other people to hone their musicianship.

Writing is a skill that gets better with practice, so start now. Get feedback later. Then write some more. In short:

JUST
FREAKIN'
WRITE.

22

Is failure necessary?

THE OVERARCHING LESSON THAT MOST STUDENTS LEARN IN THEIR MANY years of going to school is that mistakes are bad. The students who get the highest grades in school are the ones who make the fewest mistakes. Misspell a word on a spelling test, and you get points taken off. Make a calculation error in math, and you get points taken off. The more mistakes you make, the worse your evaluation. Because many people are competing for good grades to get into the best colleges, there is a lot of pressure to minimize the number of mistakes you make.

Psychologically, though, making mistakes can be a great thing.

Before we get into that, though, Bob wants to point out that the chapter title is about failure, and now we're talking about mistakes. Full-blown failure means that you had some big goal you wanted to achieve (passing a class, running a company, getting a promotion) that you did not actually achieve.

The more important the goal was to you, the bigger the failure, and the worse you feel about it. Failures are difficult to handle emotionally because they can undermine how you feel about your-self. They can lead you to believe that you are pursuing the wrong kinds of goals. You may also feel like you have let other people down.

But this chapter is going to talk a lot about mistakes for a rea-son. The consequences of the mistake determine whether it is clas-sified as a failure. When there are minor consequences, you don't think of it as a failure. When there are major consequences, you do.

You make many different kinds of mistakes, some of which have only minor consequences. You might transpose two numbers when dialing the telephone and mistakenly call the pizza parlor instead of the repair shop. Other mistakes may have minor con-sequences that could have been major. You might rush out of your house one day, fail to look both ways as you cross the street, and nearly get hit by a car. In the end, the mistake didn't have dire con-sequences, but it could have. And of course, some mistakes end up having huge consequences. You might look down at your cell phone while driving, crash your car, and injure a passenger.

For many of the complex goals in a person's life, failures are not typically the result of a single mistake. A great employee does not generally get fired for a single error. A marriage does not usu-ally split up because of one fight. A company does not typically fail because its leader makes a single bad decision. Instead, big failures often involve a cascade of mistakes that ultimately lead to devas-tating consequences. The best way to avoid big failures, in fact, is to learn from smaller mistakes and take steps to correct them before they cause bigger problems.

Now, you might be thinking: Wouldn't it be better to learn from successes? Seems reasonable enough, but successes generally don't teach as much as failures do, because it's usually difficult to diag-nose exactly *what went right* that led to the success. Was it the effort of the people involved? Was it the situation (or what psychologists often call the *context*)? Was it blind luck? Was it the great plan that was put in place?

Because it is hard to know precisely why things succeed, people often repeat what they did in the past when something went well in the hope that a repetition of the same procedure will produce the same outcome. Art frequently points to the example of Ron Johnson, who worked for Apple and oversaw the creation of the Apple Store®: the hip, sleek place to buy computers, iPads®, and iPhones®. The Apple Store concept has been lauded for its elegant design and for the way it promotes Apple's image. The Apple stores we have seen are always crowded with shoppers.

On the basis of his success, Ron Johnson was hired by JCPenney® in 2011 to be the company's CEO. His mission was to revitalize a store mostly known for selling low- to midpriced clothing and items for the home. Essentially, Johnson carried over his solution for the Apple stores to JCPenney. His aim was to appeal to a younger and hipper crowd. He changed the design of the store, aiming to make JCPenney a destination, and in the process eliminated coupons and the steep discounts JCPenney was known for. Unfortunately, this strategy was a complete failure.

Johnson's failure is easier to understand than his success. By redesigning the JCPenney stores and eliminating coupons and discounted prices, he alienated the loyal customer base. Typical JCPenney customers did not want a sleek and hip shopping experience; they were looking for a good value for their money. The new design also failed to make the store more attractive to young shoppers, who were more likely to go to retailers like American Apparel or Abercrombie & Fitch®. As a result, sales at JCPenney stores tanked, and in the end Ron Johnson was fired. Art is pretty sure that Ron Johnson learned a lot of lessons from this failure, like the value of studying a store's customer base rather than assuming a strategy that worked for a hip and cool technology company would also work for a discount clothing retailer.

IN ORDER TO LEARN FROM MISTAKES, THOUGH, IT IS IMPORTANT TO FACE UP TO them. No one enjoys the negative emotions associated with mistakes, particularly when those negative feelings are directed inward. But people, to varying degrees, have learned ways of making themselves feel better in response to errors and failures.

One way to feel better about your mistakes is to absolve yourself of responsibility and blame mistakes on the situation. By *externalizing* the problem, you avoid having to question your abilities. You did everything right, but the circumstances or the efforts of others were the source of the problem. If only things had been different, your actions would have been hailed as a huge success.

A second way to feel better about your mistakes is to ignore them. In some cases, people just stop thinking about the mistakes they make. In other cases, though, they minimize the severity of the mistakes (or are unaware of them altogether). For example, the psychologists David Dunning and Justin Kruger demonstrated that the worst performers on a given task inflate their beliefs about their performance the most. It may be that the worst performers have trouble evaluating just how bad they are, but by remaining blissfully ignorant of how badly they perform, they also protect themselves from recognizing their mistakes.

Even when people are able to recognize that they, indeed, made mistakes, they may thereafter simply avoid similar situations in which they might make similar mistakes as a way to protect their self-esteem. But developing new expertise requires trying things you have never done before. People are quite likely to make a lot of mistakes while they are learning something new, but that's no reason to avoid it.

Avoiding the possibility of failure leads you away from trying creative new ideas. Creative ideas are untested, of course, so they may not work well when they are initially tried. After James Dyson came up with his innovative idea for eliminating the bag in a vacuum cleaner, he spent years developing prototypes before his final version was ready to go to market. Along the way, he made a lot of mistakes, but he used what he learned from the mistakes to generate better designs. In the immortal words of inventor Thomas Edison, "I have not failed 10,000 times. I've successfully found 10,000 ways that will not work." (Art is a big fan of Edison, having grown up only a few blocks from the place where Edison first strung up electric streetlights in a town in New Jersey that is now called Edison.)

Instead of ignoring mistakes to boost self-esteem, it's much healthier and more productive to honestly assess what you do, accept the mistakes you make, and maintain a sense of self-compassion. Everyone makes mistakes. Only after you identify mistakes and acknowledge the role you played in them are learning and improvement possible.

Bob talks often about expert musicians who listen to recordings of their own playing to identify mistakes. Even though the musicians are highly proficient, they still spend time listening critically in order to continually improve. In fact, their expertise is a direct result of their willingness to be critical of their own performances.

THE IMPORTANCE OF LEARNING FROM MISTAKES IS ILLUSTRATED IN THE WORK of Carol Dweck and her colleagues. If you believe that your abilities are fixed (you're good at math, or bad at sports), then the mistakes you make are potential threats to your self-esteem. However, Dweck's research points out that if you believe that any ability is a skill to be learned, then mistakes are just a sign that you need to work harder. Taking this incremental skill-based approach allows you to *use mistakes* as a tool rather than fearing them as a sign of incompetence.

The acceptance of failure and the importance of learning from mistakes is also a characteristic of regions like Silicon Valley that churn out lots of new companies. Sociological studies of Silicon Valley entrepreneurs find that people who started up companies that failed were not punished for those failures. Instead, they were often brought on to work with new companies. The belief was that failure helped people learn how to be better at developing a new technology for the market. As a result, Silicon Valley created a healthy environment for developing innovative businesses. This attitude contrasted with what happened with many large high-tech companies on the East Coast that tended to punish the managers of failed projects rather than encouraging them to get involved in new ventures.

Obviously, nobody wants to fail. And history generally doesn't pay attention to ideas and products that fail. But the people who are most successful are the ones who try new things, make mistakes, and learn from them. Failure may be bad in the moment, but it can be a springboard to great things in the long run. And the people that history does pay attention to learned a lot from their mistakes.

Here's a good thing to remember:

I succeed

BECAUSE

I fail

(and learn

from it).

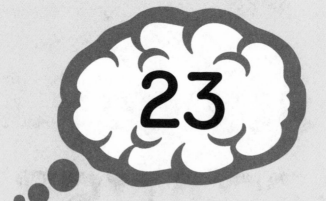

How much
of what
we see is real?

OKAY, OF ALL THE QUESTIONS IN THIS BOOK SO FAR, THIS ONE SEEMS to be the silliest. You open your eyes and look around, and you see lots of things. When we are in the studio, we see microphones, cables, coffee mugs (lots of coffee mugs), pads of paper, cups of water, chairs, pens, and pencils. It doesn't take any effort to see things. We just open our eyes, and the brain does the rest.

The world we see looks pretty orderly. That is, even when the environment is a mess (like the studio whenever the two of us are there), it is still fairly easy to pick out the objects it contains. If you look more carefully at the world, though, you can begin to appreciate how much the visual system has to accomplish to figure out where the objects are.

Suppose a pencil on the studio table is partially obscured by a sheet of paper. If you look at it closely, the yellow pencil disappears behind the sheet of paper and then reappears on the other side.

That seems obvious, but how does your visual system know that the pencil is one object, while the paper is another object, and that the pencil is passing behind the paper? That is, how does the brain lump together the elements that are part of one object and keep them mentally separate from the elements that are part of another? It performs this task so seamlessly, you don't even notice that this is a potentially difficult problem to solve.

The group of German psychologists who first began to get interested in this question in the 1930s were called Gestalt psychologists, using the German word for "shape." The focus of their research was on properties of images you see that lead you to group elements of images together into a single object. We do not perceive the individual elements of objects (lines, planes, angles, and colors); instead, we perceive objects as whole, complete forms.

For example, a key principle the Gestalt psychologists discovered is called *good continuation*. Take a look at the two simple pictures here. In the drawing on the left, the line (suppose that is a pencil) gets hidden by a rectangle (perhaps a piece of paper). Because the line seems to continue straight under the rectangle, the brain interprets the image as showing a single object that is obscured by the rectangle. But on the right, the lines don't demonstrate good continuation, so the brain assumes that it is probably two separate objects that are covered by the rectangle:

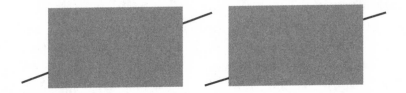

Another principle is that items of the same color are more likely to be seen as part of the same object than items of different colors. So if you see a pencil hidden partly by a sheet of paper, not only does it display good continuation, but the yellow of both ends of the pencil also suggests that the parts on both sides of the paper are from the same object.

Likewise, elements in the world that are near each other are more likely to be grouped into the same object than elements that are far apart. If you think about the picture on page 151, the narrower the rectangle, the more likely it is that the two halves of the line will be seen as part of the same object.

A fun principle (or at least one that we think is fun) is called common fate. Basically, when two parts of your visual world are part of the same object, they will also move together. If I pick up a pencil, I see all the parts of the pencil move in the same way.

Because the parts of an object that move together are generally part of the same object, when we see things moving in the same way, we group them together. Marching bands take advantage of this principle all the time when creating shows. There are fantastic videos on YouTube of college bands creating amazing displays on the field.

The members of the band line up on the field in ways that make them look like particular objects (perhaps a whale and a ship). The Gestalt principle of proximity is used to group together the individual people into clumps that can be recognized. Then, as the music plays, the marchers start to walk to particular locations on the field. In this way, the objects on the field are set in motion. The common movements of groups of musicians lead people in the audience to see those groups as part of the same object. If the marchers in the ship and the whale move in different directions, then they are clearly seen as being in different objects. Indeed, some bands march in abstract patterns, but the commonalities in movements of groups of musicians let the audience see shapes made by groups of moving band members.

These Gestalt principles are just some of the many assumptions that are built into the visual system. The behavior of light has been the same since the dawn of the universe, and for millions of years evolutionary processes have adapted the visual system to the needs of organisms. As a result, the visual system is hardwired with a number of *assumptions* about the visual world that allow the brain to make very good *guesses* about what is out there in the environment.

OF COURSE, EVERY ONCE IN A WHILE, A PARTICULAR VIEW OF AN OBJECT IS unexpected and violates the expectations of the visual system. In that case, we get optical illusions.

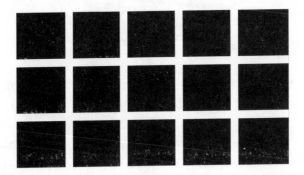

One classic example of an optical illusion is what happens with the set of black squares (as shown above). If you stare at this set of squares for a moment, you may begin to see what appear to be dark circles at the corners between the squares. If you cover the black squares, though, you will see that the dark circles are an illusion.

This illusion emerges from the way the visual system enhances borders between objects. In order to make the contour between objects as easy to see as possible, the visual system enhances the boundary between any light and dark regions of the image. For this reason, the spaces in between the sides of the dark squares look very light and bright, while the spaces at the corners of these squares, which are not bounded by as much darkness, appear darker by comparison.

The figure below shows how the *contrast illusion* can create the appearance of an object that is not there at all. The edges you see between the circles are called *illusory contours*.

When the openings of the partial circles are aligned, then the visual system uses good continuation to assume that this image is of a white square on top of four orange circles. In order to create the sense of that white square, the edges of that square are enhanced by the contrast illusion. As a result, you see the contour of a square that is not really on the page. If you cover over the circles, the edge of the square disappears. You can also get illusory contours for other shapes, like triangles.

Gestalt principles were understood by painters long before psychologists characterized them. In their efforts to depict a three-dimensional world on a two-dimensional surface, artists found ways to trick viewers into seeing things that aren't there. They exploited principles of the visual system to tease three dimensions out of two-dimensional images. At first, artists employed these principles to create more realistic depictions of visual scenes. Later they used the same principles to create images that distorted reality in interesting ways.

These interesting illusions reflect the fact that the visual system is set up to *quickly* provide a picture of your surroundings. Guessing that is based on good assumptions is necessary for making the kinds of quick judgments that allowed our ancestors to survive some pretty tough situations.

Of course, making assumptions sometimes leads to misperceptions, but there are usually multiple sources of information about what objects are, their boundaries, and their location in space. Even though Gestalt principles (or other assumptions built into the visual system) may be mistaken about what is in the world at a given moment, it is rare that a scene or object violates several different assumptions at the same time. This explains why we are rarely deeply mistaken about what we see, and why optical illusions are so delightfully intriguing.

TO

ASSUME.

Does punishment work?

W HEN YOU HEAR SOMEBODY SAY THAT A GROUP IS GOING TO BE HELD accountable for their actions, that strikes an ominous tone. After all, have you ever heard about people getting a raise after being held accountable for their actions? It just feels unlikely that anyone would ever say, "Congratulations! You did such a great job that our accountability team has decided to reward you all!"

Instead, when we hold people accountable, we are generally focusing on whether they did something wrong. If they did do something wrong, they deserve punishment. But does this focus on punishment actually work?

To answer to this question, we have to return to the motivational system, which drives people to act. There are two main subcomponents to this: approach and avoidance.

The *approach system* is engaged whenever there is something desirable in the world you want to achieve, whether it is a good meal, making a new friend, or achieving a goal at work. This system leads you to focus on the desirable aspects of the world around you and to create plans to bring about positive outcomes for yourself and others.

The *avoidance system* kicks in when there is something noxious in the world you want to stay away from, like an illness, a dangerous situation, or a failure at work. When you are in avoidance mode, you become more sensitive to all of the potential threats in the world around you, and you spend a lot of time planning for the worst-case scenario in an effort to survive another day.

The threat of punishment, of course, engages the avoidance motivational system. Each day you stave off a bad outcome is another day you live to fight again. In the short term, these threats can be very motivating. Back in the 1950s, Neal Miller examined what he called *goal gradients* by studying rats. He measured the "strength" of a goal literally by connecting a rat to a device that kept the animal in place and measured how strongly the rat would pull toward something desirable or away from something undesirable. If he put a rat near something desirable (for a rat) like a piece of cheese, it would pull toward the treat. The closer the rat was to the reward, the harder it pulled. And if Miller put a rat near something undesirable, like a cat, the rat would obviously pull away. Again, the closer the threat, the stronger the pull.

What Miller discovered is that the when the rat was near the cat, it pulled harder (to avoid the cat) than it did when it was the same distance from the cheese. That is, when the threat and the reward were close by, the strength of avoidance was stronger than the strength of approach. But when the threat and the reward were far away, the opposite was true: The rat pulled harder to get toward the distant cheese than it pulled to get away from the distant cat.

This means that in the short term, sticks can be more effective than carrots. Threatening to fire or otherwise punish a group of employees for missing a looming deadline may get them to engage with their job. So if this kind of threat makes businesses a buzzing hive of activity, we ought to use it often, right?

Wrong.

The problem is that the emotional reaction to threats is *stress*, and in the long term stress has a many negative consequences. A number of hormones are released when you are stressed that can energize you in the short term (preparing you for fight or flight, for example) but dampen the effectiveness of your immune system in the long term. So people who live with long-term stress actually get sick more often than those who don't, which ultimately affects their productivity (not to mention that they feel pretty miserable).

Another problem with punishment over the long term is that the natural reaction to punishment is to run away. For example, when Art was in grad school, his entire lab would be working late into the night in the few days before a big conference deadline. Everyone wanted to make sure they got a good grade on their papers, so the group pulled hard to reach that goal. But Art also had friends in grad school whose labs were a den of constant threats. His friends' advisors would get angry if studies weren't finished on time, if the lab wasn't kept orderly, or if the students didn't come up with what they felt were good new ideas. The students in these labs were always worried that their meetings with their advisors would end in criticism or even anger, so they avoided interactions with their advisors altogether. Many of them soured on grad school and either moved to someone else's lab or ultimately dropped out of school. They ran away from the entire threatening situation.

And that is the long-term problem with punishment. Eventually, your health declines and your productivity takes a hit, or you use avoidance tactics and look for ways to escape the whole threatening situation. Those who don't escape a high-threat environment are actually demotivated to keep working. At some point, people begin to feel as if they will get punished no matter what they do. At that point, the threats don't seem to matter much.

The occasional use of punishments can be effective in energizing people who are otherwise not very motivated to pursue a goal. However, it is important to recognize that people would *prefer* to be in a situation in which they can achieve positive rewards. A workplace or school environment that uses lots of threats will focus people on trying to find a different place to spend their time.

ONE OF THE MOST IMPORTANT THINGS TO REMEMBER WHEN CONSIDERING THE use of punishment is this: When people are working under a threat, the strongest motivation is to make the potential problem go away—not necessarily to achieve a good result. When eliminating a problem is of paramount importance, the *means* to achieving that outcome tend to matter less. As a result, people are likely to cut corners or to engage in ethically suspect behavior.

Another problem with focusing on punishment is that it tends to make people hide their mistakes, and in many situations the biggest catastrophes arise from the accumulation of small errors. These small mistakes can only be fixed if people admit to them without fear of punishment.

Indeed, one of the best examples of a culture that encourages people to admit to mistakes, without fear of punishment, is an industry where you might expect mistakes would be punished most severely: aviation. Obviously, you don't want pilots and flight attendants and plane mechanics making mistakes. Air disasters are terrifying to think about. However, the FAA has an agreement with airlines that if people report errors they make within twenty-four hours, those errors cannot affect their employment (unless they broke the law in some way). That is, for people who fly and maintain planes, there is no punishment for making mistakes, as long as the mistakes are reported.

The reason there is no punishment is that the FAA needs to analyze all of the errors that people make in order to fix any problems with procedures or with aircraft design or maintenance before the problems become part of a larger cascade of errors that leads to a disaster. Because of procedures like this, air travel has remained remarkably safe over the past fifty years, despite an enormous increase in the number of people who fly.

The healthiest work environments are ones that have a delicate balance between threats that need to be avoided (such as potential negative employee evaluations) and rewards to be approached (like promotions, raises, or access to interesting projects). In these workplaces, people are motivated to work hard enough to ensure that small problems don't become big ones, but they are also encouraged to excel.

Rewards encourage people to work hard over the long term. The hardest-working students are the ones who see their education as enjoyable and as a means to obtaining things they want, rather than as a means of avoiding failure or poverty. Likewise, the most engaged employees in any workplace are those who focus on the rewarding aspects of their jobs, like being part of a bigger mission that connects them to others and to society. The most successful and driven people—the ones who see their life's work as a *calling*— engage their approach system instead of their avoidance system.

Many of Bob's students are future music teachers. The very best teachers he knows are the ones who see their work as a mission. These teachers know the joy that playing an instrument or singing can bring to students, as well as the valuable life lessons students learn when they have to perfect their performance of a particular piece. And they also know that by cultivating musical ability and exploration, they are making the world more vibrant and joyful.

A career is a calling when the big-picture reward outweighs the threat of failure. Perhaps you are helping society through your engagement with colleagues and customers and clients. You are part of an enterprise that is larger than yourself, which helps put the occasional angry customer or stupid mistake in perspective. Almost all jobs have some threatening elements, but if there are plenty of rewarding elements to compete with them, workers will be happier and more productive. A researcher may be dedicated to curing a dreaded disease, for example. A firefighter may be focused on protecting the public from harm. A programmer may focus on protecting the personal data of clients. There are very clear threats in these professions that anyone would want to avoid, but if workers can focus on the deeper purpose of the job—the greater reward—they will ultimately do a lot more good for themselves and others than the workers who are just trying to save face and keep up with the mortgage payments.

Unfortunately, many people are unable to see their jobs as purposeful because the culture creates unnecessary threats. They work under a threat of losing their jobs if their performance evaluations are not up to snuff. Colleagues may create an unpleasant (or worse yet, unsafe) environment that makes it hard for people to engage positively with them. You have probably known colleagues who, for example, look for little ways to punish others for minor infractions. These threats undermine not just the employee but the work and the enterprise itself.

In that environment, people often focus themselves on the oasis of the weekend. The workweek becomes something to be survived rather than something to be savored. Once this happens, businesses cannot expect that their employees are giving their full effort on behalf of the organization.

In that spirit:

SPARE THE ROD

Why are comparisons so helpful?

THE STORY WE TOLD AT THE VERY BEGINNING OF THE BOOK ABOUT HOW *Two Guys on Your Head* came into being was slightly abridged. Actually, when our intrepid producer Rebecca McInroy asked us if we were interested in doing a show, we didn't say "sure" right away.

In fact, we started off by staring at her in confusion. After all, as a couple of academic psychologists, we didn't have any real experience with radio. Why in the world would anyone want to listen to the two of us?

And then Rebecca made one simple comparison that changed everything: "It will be like *Car Talk* for the mind."

Instantly, we got it. We had been fans of *Car Talk* for years. Tom and Ray Magliozzi (or Click and Clack, as they are affectionately called) had a beautiful rapport on the radio. Even people who didn't care at all about cars listened to their show, just to hang out with the two of them for an hour each week.

Anyway, it was only after that simple, helpful comparison that we looked at each other and said, "Sure." Why was the comparison so effective in changing our minds?

Whenever you try to learn something new, it is helpful to be able to attach it to something you know about already. In fact, your memory works best when the things you know are deeply interconnected. Comparing a new concept or idea to something in your memory that you understand quite well allows you to draw parallels that help you grasp the new concept more quickly and easily.

So when Rebecca told us to think of the new show as *Car Talk* for the mind, we understood that she was responding to the energy of the conversations we had, but that our conversations would be about the way people think rather than about 1979 Toyotas® with slipping transmissions. Similar format, different topic.

Here's how comparisons like this work. You start by finding points of commonality between whatever it is you're comparing—in this case, two people having a conversation about a topic. Recognizing the commonalities can have an important influence on how you think about new things. Comparing our new venture to *Car Talk* emphasized the camaraderie and (our pitiful excuses for) humor, and that comparison undoubtedly influenced the way we approached doing the show, particularly when we first started.

When it comes to finding the *differences* between things you're comparing, something interesting happens. You don't simply generate a list of all of the aspects of each thing that are not present in the other. Instead, you identify what are referred to as *alignable differences*, which, as the term suggests, are similar in kind. Having identified commonalities between the proposed new show and *Car Talk* (they both involve two people talking; one of the hosts on each show is bald; there's a lot of laughing), your mind turns to alignable differences (regarding topic: *Car Talk* is about cars, while our show is about the mind; regarding profits: *Car Talk* makes lots of money, and ours . . . not so much; regarding time: *Car Talk* is a one-hour show, while ours is seven and a half minutes).

Nonalignable differences, in contrast, are unique aspects of one thing that have no correspondence at all to the other (for instance, *Car Talk* includes puzzles that the hosts solve, while our show does not; our producer talks on the show, but Tom and Ray are the only voices on *Car Talk*). The nonalignable differences are not usually the first things that arise in head-to-head comparisons, but when they are considered, they can aid in predicting or influencing future behavior. The fact that *Car Talk* had sponsors, for example, prompted us to find companies that might want to sponsor our show (which eventually happened).

COMPARISONS USING ALIGNABLE DIFFERENCES HELP US MAKE DECISIONS without having to become experts in a particular domain. For example, every few years Art buys a new laptop. He has stopped trying to keep up with the latest processor speeds, memory capacities, and hard-drive storage capabilities. They change too often to bother paying attention. Instead, when it comes time to buy a new computer, he goes to a store and compares the models with one another.

Is a "4-gigahertz quad-core processor" a good thing? It certainly sounds impressive, although Bob prefers to measure computer performance in terms of *teraflops*, which sounds a bit like diving into a bed of flowers. Art could do a lot of research to find out whether a 4-gigahertz quad-core processor is good, but it might be easier to compare processors among the available models. Since all of the computers have processors with varying speeds, these are alignable differences. Art doesn't need to know much about processors to recognize that a computer with a 3.4-GHz processor is slower than one with a 4-GHz processor, which is slower than one with a

5.5-GHz processor. The fact that processing speeds all line up along one continuum facilitates making a comparison.

Focusing only on alignable differences can lead you to ignore some important features, though. Nonalignable differences can also contribute to meaningful comparisons, but interpreting them often requires more expertise. In Art's comparisons among competing computer models, he can see immediately that a 2-terabyte hard drive is bigger than a 1-terabyte drive. But only one of the computers Art's thinking about comes with an SD card reader. To compare computers with and without SD card readers (a nonalignable difference), Art actually has to know something about what an SD card reader is for, and whether having one would matter.

People engage in the same kinds of comparisons in social situations, too, and often without much conscious awareness. When you meet new people, you quickly have to figure out how to treat them. Do they like jokes? Are they stodgy? Do they share your political beliefs? Because you can't really ask them questions about everything, you tend to associate your new acquaintances with qualities of people you know who seem, for whatever reason, similar to them. If Bob were to meet someone who reminds him of Rebecca McInroy, for example, he may assume that she, too, is quick-witted and curious. As a result, he may answer a question she poses with more detail than she really wants to hear.

MOST OF THE COMPARISONS YOU MAKE INVOLVE THINGS THAT ARE FAIRLY similar overall—comparing computers to each other, or one person to another. Sometimes, though, you make comparisons involving things that are more distantly related. Suppose that you travel a lot, as the two of us do. It can be hard to stay in shape when you're on the road because most hotels have fitness centers consisting of four barely functioning treadmills and a TV set. It would be great to be able to bring a set of weights with you to work out, but it is cumbersome to carry weights in your suitcase (and airlines don't exactly encourage it).

A designer trying to develop a product to solve this problem might think about air mattresses, even though the air mattress doesn't look a lot like a weight set. Why on earth would someone make such a comparison?

It is cumbersome to travel with a mattress, but air mattresses solve this problem by removing the bulkiness from the mattress and then adding it back with something readily available (namely, air) when we need to use it. Likewise, it might be possible to remove the weight from a weight set and then add it back with something easily available (like water).

In fact, the history of invention is filled with stories of people using analogies like this to help them design new products. Velcro® was designed by analogy to cockleburs sticking to a dog's fur. Barbed wire was designed by analogy to the ocotillo plant found in the American Southwest. James Dyson developed his famous bagless vacuum by analogy to the industrial cyclones found in sawmills.

What makes analogies so powerful is that people are remarkably good at drawing comparisons between items that are not similar on the surface but share what are called *relational similarities*. Air mattresses and water weights don't look similar, and they don't involve the same elements, but they *do* share the relationship of removing a cumbersome element from the product and replacing it at the point of use with something easily available. This essence forms the basis of the analogy.

All of this work suggests that when you are trying to learn something new or to solve a new problem, the best thing you can do is to ask, *What is this like?* Then let your ability to make comparisons between new things and what you already know take over.

Analogy IS THE MOTHER OF invention.

Why do people choke under pressure?

ART USED TO WATCH GOLF ON TV. HE IS NOT TOO PROUD TO ADMIT IT now. Golf is a strange thing to watch on television—you spend a lot of time trying to pick out the ball against the blue sky and listening to the hushed tones of the game's commentators. But this pastime gave Art a chance to witness one of the biggest collapses by a player in the history of sports. In 1996, Australian golfer Greg Norman entered the last round of the Masters® (one of professional golf's biggest tournaments) with a commanding lead. Over the last eighteen holes, however, Norman played horribly, and he ultimately lost the tournament.

When a normally skilled individual has a terrible performance in a high-stress environment, we call that *choking* under pressure. One of Art's graduate students, Darrell Worthy, analyzed the game transcripts from every NBA basketball game played over three seasons, specifically looking at the free throws shot in the last minute of games in which the score was close. He compared each player's percentage of shots made in the last minute to the same player's season average for free throws. His analysis revealed that when a player's team was down by one shot (so his free throw could tie the game in the last minute), the player shot several percentage points worse than his season average. When the game was tied (so his free throw couldn't hurt the team but could certainly help it), he shot a little bit better than his season average. It seems that even NBA ballplayers choke under pressure.

There are many different forms of choking. Athletes may suddenly play poorly in key situations. Students may get a bad grade on a test they thought they were well prepared for. Speakers may lose their place in front of an audience. Actors may forget their lines when they step out onstage.

Stress does not always lead to poor performance, though. Baseball player Reggie Jackson used to be called Mr. October because of his heroics in play-off baseball. In 2015, Daniel Murphy, a New York Mets second baseman who had hit only fourteen home runs the entire season, hit home runs in six consecutive play-off games to help send his team to the World Series. Chicago Bulls legend Michael Jordan routinely sank amazing shots to win basketball games, even in clutch play-off situations.

As these examples make clear, performance pressure doesn't always lead to choking. Indeed, some people seem to excel when the pressure is high.

Why is that?

PERFORMANCE PRESSURE HAS TWO DISTINCT INFLUENCES ON THE BRAIN.

First, pressure decreases your working memory capacity. As we discussed in earlier chapters, working memory is the amount of information that you can hold in mind at any given moment. The more information you can hold in mind, the more you can make connections among pieces of knowledge in ways that allow for more creativity and flexibility in your behavior.

But when you're stressed, the capacity to hold various pieces of information in mind decreases. This may seem like a bad way for your system to be set up, but when you are under threat of attack (a major source of stress for our evolutionary ancestors), it is best to focus on the most important information available and move rather than take time to think creatively about the variety of ways you might escape. The brain's reactions to stress help you to find a way out of a dangerous situation quickly.

Second, pressure causes you to pay more attention to your own behavior. If you are in a stressful situation, then it is possible that a wrong move could lead to failure. Indeed, in our evolutionary history, many stressful situations were life threatening. Under those circumstances, it makes sense that stress would lead you to pay more attention to what you're doing.

The modern world, though, is quite different from the Paleolithic world in which our ancestors lived. Most of the stresses in our contemporary industrialized societies don't come from creatures or other people that want to snack on us or capture our territory. Instead, they come from social situations, the expectations of bosses, and sporting events. We often become deeply engaged in these non-life-threatening situations and can feel powerful stress from them. But responses that evolved over millennia are not always well adapted to the modern world.

And that can lead to choking under pressure.

BOB GETS TO WATCH A LOT OF MUSIC PERFORMANCES AS A PROFESSOR IN A music school. At times, music students whose performances are being graded find themselves unable to successfully perform a piece they had practiced for months. Why this happens is particularly interesting because it involves shifts in the way people think and pay attention.

Extensive practice trains the brain to execute highly refined, skilled movements automatically, without conscious control. In a pressure-packed evaluation situation, however, the powerful frontal lobes of the brain begin to monitor these well-practiced movements, disrupting the timing and coordination of the signals that control them. Suddenly a piece that sounded so beautiful in the practice room is riddled with flaws.

This kind of monitoring happens in many situations. When Art was in high school, he was in a few plays. He distinctly remembers a performance in which he suddenly became aware of his hands. Normally, you move your arms and hands in an automatic way, without having to give attention to which limb is doing what. The stress of being onstage led Art to start thinking about his own movements, and he suddenly moved awkwardly across the stage as he tried to think explicitly about what to do with his hands.

Try it yourself. You probably have no idea what your natural hand movements are like, so when you try to *decide* how to move your hands, you end up looking and feeling rather strange.

There are things you can do to help guard against the problems caused by self-monitoring. When you practice a skill that involves body movements (like playing an instrument or a sport), you should also practice *what you will focus on* when you execute the skill in a pressure situation. That way, you develop a habit of focusing on something specific and are less likely to shift your attention to monitoring your own movements.

Baseball players, for example, may practice thinking about the game situation as they practice hitting. When they are feeling performance pressure during an at-bat, they can concentrate more on elements of the game rather than being hyperaware of their execution of the swing.

THE DECREASE IN WORKING MEMORY CAPACITY THAT WE MENTIONED EARLIER causes problems in situations in which you need to think expansively or creatively, and it is one of the reasons why test anxiety is such a challenge for many students. Tests often require flexible thinking, and students who get very nervous in situations in which they are being evaluated find it difficult to think flexibly because their working memory capacity is diminished as a result of the anxiety. For this reason, their performance on big tests seldom reflects the extent of their knowledge.

Quite a bit of research on academic stress has been devoted to math anxiety. For many years, educators have noticed that boys are much more likely than girls to focus on math and to seek careers in math as they get older. Psychologists have observed that the average level of anxiety about doing math is higher for girls than for boys, particularly once they reach middle school.

A fascinating thing about math anxiety, though, is that girls experiencing math anxiety are more worried about whether they are good at math than about particular tests they are taking. So while boys and girls actually experience about the same amount of anxiety while they are taking exams, girls are more worried that they should be doing better in math than they are.

Female underperformance on math tests might be the result of a type of pressure called the *stereotype threat*. Stereotype threat occurs when you are part of a group that is associated with a negative stereotype. When you are in a situation that may confirm the stereotype, you experience additional pressure that may harm your performance.

For example, in the United States, there is a pervasive

stereotype that African Americans are not as intelligent as Whites. African Americans may underperform on a formal test of intelligence when they are concerned that their performance may reflect badly on the group to which they belong. This stereotype threat effect has been observed in many different situations. For example, White men do worse on math tests when they are in a room full of Asian males than in a room full of White males because of the common stereotype that Asians are better at math than Whites are. Of course, no individual consciously acknowledges these effects on his performance, and often the sizes of the effects are small and inconsequential. Nevertheless, they illustrate how our own expectations—and the pressures associated with those expectations—can affect our performance.

We'll close this discussion with the suggestion that the best way to handle performance pressure is to actually practice by putting yourself in pressure-packed situations—even those that you create for yourself. The more often you face situations that scare you, the more opportunities your brain has to learn that the threat is not as daunting as you might imagine. Your performance in a single instance seldom has lifelong consequences. After many experiences performing under pressure, your motivational system learns that the stress response is unnecessary or even counterproductive. You learn to keep your cool.

PRACTICE
WITH
PRESSURE
PROMOTES
POISED
PERFORMANCE.

How do we decide what to buy?

A COUPLE OF YEARS AGO, ART HAD TO BUY A NEW COFFEEMAKER. HE went to a local Bed Bath & Beyond®, a cavernous retailer of things for the home, where he was confronted with an entire section devoted to coffeemakers. The array of features was dizzying: cone filter, basket, or K-Cup®; different sizes; automatic timer; glass or metal carafe. There must have been over thirty to choose from.

There were so many options, it quickly became clear that Art was not going to have a peak coffeemaker experience. He nearly ran out of the store screaming. There was simply no way to figure out which model was the best one for him—at least not without a detailed guide explaining which features mattered and a few days to consider all of the options carefully, or perhaps a coffeemaker concierge to provide assistance and advice.

Ultimately, Art grabbed a twelve-cup coffeemaker with a cone filter that looked like it probably wouldn't fall apart too quickly. He recognized the brand name, so he assumed it was probably reasonably well made. He spent about ten minutes figuring out what to buy. After using his purchase for several years, Art is happy to report that the coffeemaker still makes a pretty decent cup of coffee, and his decision seems to have been a good one.

This example says a lot about how we normally shop.

Most of the products we buy are things we know very little about. We may have heard the names of a few of the manufacturers, and we may have in mind a couple of features that matter to us, but a lot of the subtle differences among products are not things most people know about or even care about.

So how do we figure out what to buy?

The explanation offered by economic theories of decision making suggests that we make choices by considering each of the features of a product (including price), figuring out how useful the features are, assigning relative weights to the features in terms of their importance, and adding up the total usefulness (what economists call *utility*). The item with the highest utility is the one we choose—in *theory*. It's a tidy, seemingly reasonable explanation, but in most cases it's not how things actually work.

In most circumstances, evaluating each aspect of a given product and weighing the features against its cost is far too time consuming for us humans to carry out—especially in today's crowded, fast-paced world. Most of us have neither sufficient knowledge nor patience to evaluate each option carefully. Instead, we use the readily accessible information in our environment and in our memories.

And one influential and readily accessible piece of information is how familiar a product or brand name seems to us in the moment. In fact, familiarity is one of the primary drivers of purchase decisions. We are all wired (that is, innately predisposed) to mistrust things that are new and unfamiliar and to like things that are familiar. Psychologist Bob Zajonc (pronounced ZI-*ons*) did studies demonstrating what he called the *mere exposure effect*. Just showing something to someone (exposing them to it) made it more attractive later on.

Melanie Dempsey and Andrew Mitchell showed that this effect is present even when people don't realize that they've been exposed to something before. In one of their studies, participants were shown images of a pen paired with some other positive information, but the images were flashed on a screen so quickly that it was impossible for the viewers to consciously recognize them.

Later, the participants had a chance to select from a set of pens, including the one that had been flashed subliminally (below conscious awareness) earlier in the study. And guess what? People tended to choose the pen they had "seen," even when there were other (unfamiliar) pens in the set with features that made them objectively superior to the familiar pen. Something as simple as the flash of an image was enough to make one pen seem more familiar than (and preferable to) the other pens, even when participants were not consciously aware of having seen the pen before. This suggests that even if we try to ignore the ads in our environment, they may still influence our decision making, and it helps explain why Art ended up buying a coffeemaker from a brand he had heard of before.

IF YOU HAVE EVER BEEN TO A ROCK CONCERT, YOU HAVE PROBABLY EXPERIenced the mere exposure effect. The headliner band plays a two-hour set, and they save their biggest hits for the end of the show. The songs that get the biggest reaction from the crowd are the ones that have been played often on the radio. While it is possible that those are objectively the band's best songs, chances are that is not what people are responding to. Instead, the big reaction of the crowd happens because those songs are familiar—and that familiarity translates into preference.

Companies spend a lot of time, effort, and money making their products seem familiar to you, and by "familiar" we mean "easy to recall from memory." In fact, that is one of the most important functions of advertising. Sure, ads often have images of satisfied customers, and they sometimes provide information about how and why the product works as well as it does. But the critical thing an ad does is make the product feel familiar, so that it's more liked when it's seen at the store.

Advertising and familiarity are not the only factors that affect our daily choices, of course. We also use information determined by the set of options that are available to us in the moment. Next time you're in an upscale restaurant, take a look at the wine list and ask yourself what leads you to choose the wine you do. Most wine lists include some relatively inexpensive wines, some absurdly expensive bottles, and other choices distributed in the middle of the price range. This arrangement of choices isn't an accident. Marketers and sommeliers understand very well another aspect of decision making, called the *compromise effect*. When making decisions among varying prices, we most often choose items toward the middle of the price range. We don't typically buy the cheapest wine on the list. And most of us certainly don't buy the $1,200 Châteauneuf-du-Pape. Instead we tend to buy wines that are not too expensive but not too cheap, either. The wine we eventually choose is a function of not only the wine itself, but also the presence of other wines on the list.

In most purchase decisions, there's a trade-off between price and quality. The highest-priced items are usually—though not always—the highest-quality items as well. And when we don't know a lot about the products we're considering, we often use that trade-off to decide which item to buy. Usually, we don't want the lowest-quality item, because we are afraid that we might not enjoy it or it won't last, but we don't really need or might not have enough money to buy the very best-quality "luxury" item, either, so we tend to pick the one in the middle.

Perhaps the most amazing thing about the choices we regularly make is how little time we spend on them. It turns out that people often make a trade-off between effort and accuracy. That is, you aim to spend the least amount of time on a choice that will allow you to make a choice that's good enough for the situation.

Think about what happens when you go to your local drugstore. As you are standing in front of the register paying for your items, you think you might want to buy a candy bar. There are probably forty or more candy bars to choose from. You could analyze each one carefully and make a reasoned choice (although you may annoy the people behind you in line). Instead, you probably select something after a couple of seconds of looking.

Part of the reason you can make the choice so quickly is that there is not much danger in making a choice that isn't perfect. Perhaps the best option for you at that moment is a Snickers® bar. If you end up buying Reese's® Peanut Butter Cups instead, you are still probably going to enjoy your candy, even if you missed out on the ideal choice.

However, if you are allergic to peanuts, then you might spend more time reading the ingredients to make sure that you get something that will not make you sick. Because the cost of a mistake is high, you put in more effort to make sure your choice is acceptable.

One other reason we spend so little time on choices is that we have to make a lot of decisions each day. If we deliberated carefully over all of them, what are now routine aspects of our lives would take up nearly all our time. On a typical weekly trip to the grocery store, a shopper might buy fifty items. If each choice took a full minute, that would be fifty minutes just to choose the items, plus time to walk through the aisles, wait in the checkout line, and get to and from the store. That's a hefty chunk of the day. In fact, analyses of shopping behavior suggest that most of the items we select are essentially the same ones we have chosen before. Even though comparison shopping might save us a little money, chances are that the time we save by selecting items quickly outweighs the potential benefits of making more careful decisions.

All of us have developed a lot of strategies—some conscious, others not—to help us make pretty good decisions without spending too much time on them and without having to become experts in everything. For choices in which making a mistake is potentially consequential, we might consult a website or two, looking for reviews and more detailed information before making a decision. For very big purchases, like buying a car or house, we might even take several weeks to make a decision, read lots of reviews, test out each option carefully, and enlist the help of experts.

A key element in making choices, though, is what Nobel laureate Herb Simon brilliantly termed *satisficing* (a mashup of *satisfy* and *suffice*). In most instances, resources are limited. You don't have enough time, money, or energy to make the very best decision all the time, so you are generally satisfied by making decisions that are "good enough" for the situations you find yourself in.

The perfect is the enemy of the good is a well-worn adage about how insisting on the very best often gets in the way of our being satisfied and happy with things that suit our purposes perfectly well. Perhaps we could adapt that here to say:

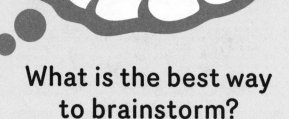

What is the best way
to brainstorm?

YOU HAVE PROBABLY BEEN PART OF BRAINSTORMING SESSIONS IN THE past. A group gets together to solve a problem. Everyone starts throwing out ideas and building on the things other people say. Lively, collaborative discussions like this can be great fun, and individuals who participate often feel as if they contribute to finding a good solution to a knotty problem.

When Alex Osborn coined the word *brainstorming* in the 1950s, he laid out the basic rules for getting groups to be creative. He wanted people in a brainstorming group to come up with as many ideas as they could, to avoid criticizing the ideas other people generated, to build on the ideas that were under discussion, and to remove any constraints on the problem in case those constraints were limiting creativity.

This set of rules seems so reasonable that you probably have never questioned them—but psychologists have.

Experiments that evaluate the effectiveness of brainstorming techniques generally compare groups who use the rules of brainstorming with identical numbers of other individuals who are asked to generate ideas working alone. The typical finding is that the individuals working alone come up with more ideas than the groups. Well, what about the quality of the ideas? Maybe the group is just more efficient, so they generate more really *good* ideas. Nope. Studies have shown that when ideas are evaluated for whether they are new and useful, the ideas generated by the individuals tend to be better than the ones generated by the group. This finding has been obtained often enough in brainstorming situations that it has been given a name: *productivity loss*.

So why does brainstorming fail? And what can you do instead if you want to become better at helping groups come up with ideas?

There are several reasons why brainstorming doesn't live up to its promise. In fact, many of the intuitively attractive rules of brainstorming actually lead to its ineffectiveness. Let's consider them one at a time, starting with the notion that getting together in a group is a good way to think creatively.

Two different processes are involved in the act of thinking creatively, one of which is referred to as *divergence*. In divergent thinking, the aim is to come up with many different possibilities. There are lots of tests of divergent thinking. Asking people to come up with as many alternative uses for a brick as they can think of, for example, tests their divergent thinking.

A second part of creativity involves evaluating the various ideas you develop and deciding which ones are best. The result of this process is to narrow the set of options being considered. This part of the creative process is often called *convergence*.

The problem with getting groups together is that the first person who says something contaminates the memory of everyone else in the room. Usually, the person in the group who most enjoys being the center of attention will speak first. (Art often volunteers his ideas quickly when working with a group.) Being the first person to speak does not mean that you have the best idea. It just means that you had the first idea that was spoken aloud.

That idea gets into the minds of everyone else in the group, and they start thinking about all the other things that the idea reminds them of. Unknowingly, Art's outburst has narrowed the thought process around the idea (thanks, Art . . .), making everyone else in the room think about the problem more similarly and hampering the divergent thinking aspect of creativity.

Indeed, lots of work on group dynamics suggests that groups are particularly good at coming to a consensus. Pairs of people talking together and groups that spend time together generally end up thinking about things in the same way. The concept of *groupthink*, often invoked in the press, refers to the idea that the more a group interacts, the more that individuals in the group view the world in the same way. That can be a great force for keeping groups cohesive, but it is bad for creativity.

All of this does not mean that we should not use groups when trying to be creative. Instead, we should recognize that when it is time to diverge, individuals in the group should actually work alone. Then the ideas should be passed around to each group member (still working alone), and everyone should have a chance to build on those ideas.

The advantage of doing it this way is that each member of the group has a chance to come up with as many ideas as they can before they are contaminated by other people's ideas, and the broad base of knowledge of the many different people in the group can be tapped as effectively as possible. Each individual also gets to expand on the ideas that other people come up with.

Only after everyone has generated ideas on their own and then built upon the ideas of others should the group meet together. The purpose of the group discussion at this point in the process is to begin to decide which ideas should be chosen as good solutions to the problem. Because groups tend to converge, this discussion helps the group reach a consensus. And because everyone ultimately contributes to the discussion, the group process helps everyone in the group to feel some ownership of the solution or solutions that are ultimately selected.

This revised brainstorming strategy helps to guard against the influence of people who really love their own ideas. Some people (particularly those who are high in the personality characteristic of narcissism that we talked about earlier in the book) want their ideas to be the chosen ones. Individuals fighting hard in favor of their own ideas can diminish the likelihood that the group comes up with what may be even better ideas.

ANOTHER RULE OF BRAINSTORMING THAT CAN HAMPER CREATIVITY IS THE ONE about ignoring constraints on a problem. It seems reasonable to assume that imposing constraints on what a solution should be may keep us from considering really novel solutions. Actually, though, when there are no constraints on a problem at all, people tend to be *less* creative. That's because the description of the problem, along with its constraints, reminds people of things they know, and those memories affect the ideas they create. It is harder to come up with viable solutions when there are constraints, but when you *do* come up with a solution, it is often quite novel.

Tom Ward has done many experiments in which he asks people to draw animals that do not exist or animals that come from alien planets. Interestingly, most of the things people draw (even when they are supposed to come from alien planets) are structured much the way people are, with separate sense organs for different senses (often the same kinds of senses that people and Earth animals have). The beings they draw are usually symmetrical, with pairs of appendages of different kinds. And the more intelligent the creatures are supposed to be, the more they tend to look like humans.

People's drawings get more creative, however, when they are given more constraints on the problem, so that some of the obvious similarities between Earth creatures and alien creatures cannot be used in the task. If the animals come from a planet in which they cannot touch the ground, for instance, then people have to develop mechanisms for them to stay afloat, and that influences their appendages and often their sensory systems as well.

In practical settings, constraints matter, because the concepts that emerge from brainstorming sessions need to be both new and useful. Real creativity typically involves finding solutions to problems that obey the constraints that make them useful.

So, next time you're under pressure to come up with a brilliant new solution, remember this:

Individuals diverge. Groups converge.

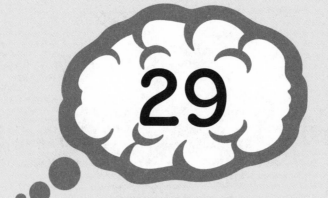

Why is online communication so ineffective?

FIVE HUNDRED YEARS AGO, ALMOST ALL HUMAN COMMUNICATION WAS face-to-face. It was possible to write notes, of course, but sending them over any distance was difficult or, in some places, nearly impossible. Fast-forward to the early twenty-first century, when instant communication around the globe is commonplace. You can pick up the phone, make a video call via the Internet, and send a text or an email to almost anyone, anywhere.

The ability to be in touch with people all over the world at almost any time is fantastic, but the forms of communication we use today are very much unlike the circumstances that were present when language evolved, with small numbers of people communicating over short distances in real time. When we communicate over long distances, we distill communication down to the basic units of language that are necessary to transmit information, but human communication comprises more than just the words that are spoken.

When Bob says, "Hey, nice shirt!" he is probably being nice. When Art says the same thing (particularly to Bob), he is probably not. The same words are spoken, but the intonation differs. Plus, the context is missing. Does the fact that Bob is wearing a garish Hawaiian shirt and that Art is wearing a stylish button-down shed some light on the meaning of those words? Probably so. The words by themselves don't convey the full, true message, and given their intonation and context, they might actually express opposite meanings.

Linguists refer to the elements beyond the words that help us communicate as the *pragmatics* of language. When two people are in conversation, their tone of voice, cadence, the rising and falling pitch of their sentences, their gestures, and their facial expressions all affect how their words are understood.

Which brings us to email. In just twenty-five years, email has gone from being something that a few high-tech people used regularly to a communication channel that is nearly universal. On a typical day, everyone in a workplace may receive fifty emails or more that are directed specifically to them, plus another fifty that are meant to provide an update, and perhaps another fifty intended to sell them something. Because email is so easy to send and arrives at its destination so rapidly, we use it for a variety of functions: We hold conversations with colleagues or friends; we write long requests to people; we complain about problems. We used to do these things in person or over the phone, but now, we often write them down and push SEND.

Before focusing on the problems that arise when so much of our communication is written, it is worth saying a word about the frequency with which emails arrive. Because it takes milliseconds for an email to arrive at its destination, almost anywhere on the planet, we often feel as though it should be answered as quickly as it is received. But with the number of emails people receive each day, doing so would keep us checking and responding to email all the time. The beeps and badges signaling that new email has arrived drive us to keep shifting our attention and checking what's coming in. That is an invitation to multitask. And as we discussed in a previous chapter, humans are lousy multitaskers.

Although some of us may work in environments where the senders of email (impatient bosses) expect almost immediate responses to their little missives, many of our emails can wait several hours before being answered. And they can even wait several hours before being *read*. Most of us would be a lot more productive if we cut down on the number of times that we check email each day, setting our software to check only at defined times—preferably during periods of limited productivity and creativity.

Now, aside from distracting us from more important matters, emails, texts, and chats can lead to the kinds of miscommunication that can have negative effects on relationships.

Take simple requests, for example. When someone asks you to do something, either in person or over the phone, the requester's tone of voice can express the urgency of the request. It can also help to express whether the request is a demand or a plea for help. Without some sense of the sender's tone of voice, it is easy for the recipient to misunderstand the significance and urgency of the request, leading to judgment errors about when and how to respond. And those of us who become overwhelmed by the flood of emails relentlessly surging into our inboxes may mistakenly interpret the presence of a new email as a need to act *right now*. This is especially problematic because responding to an email usually takes less time than does completing the report you've been working on all week. The sense of accomplishment and completion that comes with responding to an email ("Done!") is the prime reason for its seductive power. Spending more time on the report today (still not done) is weak competition for your attention.

Determining what, when, and how much effort to devote to tasks is relatively easy to negotiate in person or over the phone. This kind of back-and-forth is harder and more time-consuming via email, so people often just assume the request is urgent (and then resent the sender for initiating the request).

Another problem with email comes from the distance between the people communicating. If you're like most people, when you are in face-to-face interactions you try to be considerate, and being considerate means that you take the feelings of the person you're interacting with into account. Although there are certainly people and circumstances that seem to foster tension and arguments, most of our face-to-face interactions with other people are relatively pleasant.

In the moment, we can see the influence our words have on the people we're talking to. If you say something that is confusing, or pleasing, or hurtful, you most often can detect the emotional reaction right away. But when we send an email, we can't see the reaction to what we say as we're saying it. As a result, with our human sensitivity filter turned off, we may phrase things in ways that are terser, coarser, or nastier than we would if we were watching the reactions of our correspondents. The distance in time and space between the sender and recipient makes it easier to offend or hurt, because we can't see or *feel* the emotions of the recipient.

People who communicate often by text and chat recognized the emotional ambiguity problem, which led to the creation of *emoticons* (character-based configurations meant to convey emotion), such as :-), and *emojis* (pictorial configurations like ☺). These images are intended to add some emotional tone to written text. These and other conventions—YELLING IN ALL CAPS!, for example—certainly help to eliminate some emotional ambiguity, but reading emotional tone from hastily written text remains a challenge.

THE ABILITY TO HOLD A CONVERSATION IS ACTUALLY A SKILL. SITTING AND talking with other people in a room involves using vocal inflection, facial expressions, and gestures to communicate. Good conversationalists leave space between sentences for other people to talk. Without practice, it is hard to be an effective participant in conversations.

Listening is a crucial skill in conversation. Many people pay attention to what other people are saying just long enough to figure out what they're going to say next. It takes patience to really hear what other people are trying to communicate and to respond appropriately. When you observe and respond to the reactions of those you're conversing with, there is a give-and-take that fosters relationships.

When we communicate with others primarily through written texts, though, we miss out on the important contributions that emotions make to building and sustaining positive relationships. These emotional interactions make the tasks of life and work more pleasant.

Okay, so we're not a couple of Luddites suggesting that email, text, and chat be abandoned completely. They serve important roles and are often used quite effectively, but it's important to note that words are just a small part of human communication, and they can easily be misinterpreted. Remember:

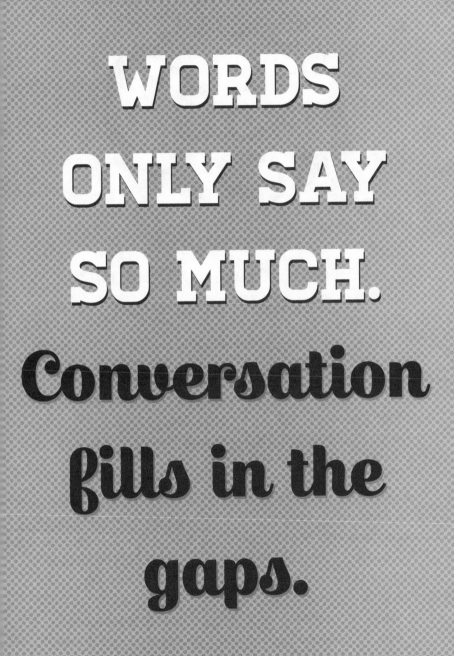

WORDS ONLY SAY SO MUCH. Conversation fills in the gaps.

Is it possible to remember something that didn't happen?

WHEN ART WAS A KID, HE HAD A STRANGE EXPERIENCE. HE WAS hanging out at a friend's house, and his friend told him a long story about an adventure he had hiking by a lake when he encountered an injured duck. He carried the duck back to the house, and the family took care of it until it was well enough to return to the lake.

The story had detail and clearly stirred up a lot of emotion in Art's friend, but just as the story finished, the friend's older brother leapt into the room, swatted his younger brother on the head, and said, "Dude, that didn't happen to you, it happened to me." A little questioning of the parents made it clear that the older brother was right. Art's friend had told a true story, but he mistook himself for the main character.

How can something like that happen? How can you possibly have a memory that seems so vivid and so emotional that you believe it deeply, even though it has key details that are deeply mistaken?

This question has been explored by psychologists for several reasons. Obviously, the idea that people might have memories that include false information is interesting in its own right. But in addition, our legal system relies on people being able to provide reliable eyewitness reports. If people's memories for events they experience are not dependably accurate, that calls into question a lot of eyewitness testimony.

Our culture really likes to assign credit and blame. We name particular individuals as the inventors of technologies, for example, even though most inventions develop in a community of people who are working on similar problems. To accurately assign credit and blame, though, we need to remember exactly who said and did certain things.

IN ORDER TO UNDERSTAND HOW WE CAN HAVE FALSE MEMORIES, IT IS IMPORTant to talk a bit more about the way the brain stores information. We often compare the brain to a computer. In fact, there is a lot that is useful in the computer metaphor, but memory storage is not one of them. In a computer, long-term data are sent to addresses on a disk or other storage media and are retrieved without errors when needed.

The brain does things differently. Information is distributed throughout the brain, and only a partial match of a pattern you encountered in the past can lead a memory to be retrieved. The cost is that the memory is really being *reconstructed* at the time it is used, and each time a memory is reconstructed, it is likely to gather parts of lots of different, similar memories.

That is where the computer metaphor breaks down. With a computer, it is critical that the document you stored is retrieved exactly as it was when you left it. The brain has a different purpose. A lot of what the brain is doing is trying to predict what is going to happen next. What is important to the brain is that the *prediction* is correct—not that you remember a previous event exactly as it happened.

An important aspect of the way the brain stores information about events from your life is that the content of what you encountered is separated from its source. The only way to match up a particular memory of a sight or sound with the situation in which you encountered it is to call up both the information and the source at the same time. If you lose some of that source information or don't retrieve it, then you may start to blend together information from different sources and assume it is all part of the same event.

Elizabeth Loftus and her colleagues have been exploring phenomena like this since the 1970s. We already talked a bit about Elizabeth Loftus in the chapter on why stories are memorable. In the study we described, the words people heard affected what they remembered later.

In her classic study, participants watched a movie of a car accident in which one car drove through a stop sign and hit another. Later, participants were asked a question that contained misinformation: "How fast was the car moving when it ran through the yield sign?" This question is focused on the speed of the car, and the false information is that it ran a yield sign rather than a stop sign.

Later, participants saw pictures and were asked which ones were part of the movie they'd seen earlier. Participants who heard the question about the yield sign often selected an image of the car from the movie driving past a yield sign rather than a stop sign. This finding suggests that people are merging information they *heard* with information they *saw*.

It turns out to be remarkably easy to create these kinds of false memories. In later studies, Loftus and her colleagues were able to convince many college students that they had been briefly abducted from a shopping mall when they were five. Essentially, people were blending their own early memories with information given to them by the experimenter, eventually filling in details that were not part of what had been told.

PEOPLE OFTEN EXPERIENCE FALSE MEMORIES WHEN THINKING BACK TO EARLIER events. They put together memories of what they experienced that combine stories told about the event and even home videos that they saw much later. Art's friend remembers seeing the duck and has seen pictures of it that were taken at the time. Over time, though, he forgot that the source of this information was watching his brother bring home the duck rather than doing it himself.

In the 1990s, Roddy Roediger and Kathleen McDermott revived a lab study technique first used in the 1950s by James Deese. In this technique, participants hear a list of fifteen words that are all related to a target word. For example, in a list where the target is *window*, the words would include *glass*, *pane*, and *sill*. The target word itself is never on the list.

Later, participants are asked to remember as many of the words on the list as they can. Depending on how the study is done, between a quarter and half of participants will recall having seen the target word, even though it was not on the list. The idea is that all of these words related to the target lead participants to think about the target word. If participants cannot distinguish between what they heard and what they merely thought about, then they falsely recall the target word.

Factors that make it difficult to distinguish sources of information increase the tendency to falsely recall something. For example, when participants engage in a mindfulness meditation technique, they focus on experiencing their thoughts rather than evaluating them. Afterward, if they are given word lists like the one described above, they falsely recall the target words more often than they would without engaging in mindfulness meditation.

Finally, the separation of source information from the content of what you know also relates to the experience of déjà vu. (This is about the point where Bob will break down and ask whether we have written about this already . . .) Think about what happens when you encounter something new. You see the new thing and you don't retrieve any memories that relate to it. As a result, the source memory system is not active. So you realize that you have never seen this thing before.

But suppose the new situation is vaguely related to something you have encountered before. Now, suddenly, you see something that you know you have never seen before, yet you feel as though it is familiar, because you have a sense of where and when you saw it. That leads to the experience of déjà vu.

People want to ascribe all sorts of mystical meaning to déjà vu experiences. They believe them to reflect past lives, supernatural powers, or even the incredible ability of dreams to predict the future. It is natural to be amazed by the seeming familiarity of something you know you have never encountered.

Your subjective experience of the world does not separate the content of memories from the source of those memories—you simply remember things you have encountered before and have some sense of where you saw them. So it doesn't generally occur to you that these aspects of memory can be stored in different spots in the brain. As Bob likes to say, most people are unaware of the fact that memories are just reconstructions of the past.

When compared with these supernatural possibilities, then, the truth is somewhat disappointing. Your source memory system just misfired and made the unfamiliar feel familiar. In other words:

Déjà vu

IS LESS

MAGICAL

THAN

IT SEEMS.

31

Is prejudice avoidable?

ART MOVED TO AUSTIN, TEXAS, FROM NEW YORK CITY. HE WAS teaching at Columbia University, which is located on the Upper West Side of Manhattan. The neighborhood around the university is highly diverse. It is about five blocks from the official boundary of Harlem. The collections of people on the street were a huge racial and ethnic mix. The signs that hung on the bus stops in the neighborhood reflected a number of different languages.

Austin was quite different in ways that were surprising. The neighborhoods were very segregated. Soon after moving to Austin, Art was invited to an event that had a catered lunch. During the event, he felt a little uncomfortable, without being able to put a finger on why. Later he realized that nearly everyone seated at the tables of this event was White, while nearly everyone serving at the event was not.

In fact, the kind of segregation Art experienced in Austin is much more common in cities and towns in the United States than the mix of different people that characterizes Manhattan's Upper West Side.

Psychologically, there are several things that perpetuate this kind of segregation.

First, there is a tendency for people to like other people who are similar to them in some way. People tend to feel more comfortable with others who have the same background, have a similar level of education, and like the same activities and events. Whether you realize it or not, you also use skin color and other markers of race and ethnicity to make quick judgments about whether you are likely to be similar to someone else.

Second, the people you spend time with end up forming an *ingroup* that naturally creates an "us" versus "them" distinction. People in your ingroup are treated differently from those outside your group (naturally referred to by psychologists as the *outgroup*). These terms have the quality of being both clear and awkward at the same time.

It turns out that there are lots of ways to characterize someone as a member of your ingroup. Personal acquaintance plays a role, for sure. Your friends are generally part of your ingroup. Membership in a common institution or activity works as well. When you are a student, the other people who go to your school are ingroup members, while those who go to rival schools belong to an outgroup. Similarly, if you work for a company, other people who also work for the company are often considered part of your ingroup, even if your organization is spread across the world.

Of course, defining your ingroup depends on how you categorize yourself at any given moment. On an ordinary day at work, Bob might think of himself first as a member of the music school and consider people in other departments to be part of the outgroup. But if there is a discussion involving The University of Texas and a rival school (like Kansas or Texas A&M), then his ingroup might extend to all of The University of Texas. And if he reads an article about politicians attacking the rights of faculty at a university in another state, the ingroup might extend to academics in general.

In experimental studies, it is even possible to create ingroup and outgroup effects using something known as the *minimal groups technique*, by assigning people arbitrary labels. For example, in some studies, experimenters have people look at a large array of dots and ask them to estimate how many dots they see. Then they give some people feedback that they overestimated how many dots there were and insist that there is something special about "overestimators." From that point on in the experiment, those people treat other overestimators as members of an ingroup and dot underestimators as members of an outgroup. (In good experimental fashion, of course, other people are told that they underestimated the number of dots. Everything works the same way.)

When someone is a member of your ingroup, you tend to focus on their finer qualities and assume that those qualities are a result of membership in the group. For example, a generous act by an ingroup member is evidence that your ingroup is a generous bunch. Negative actions by ingroup members, though, are minimized in some way. One reaction is to justify the action based on the circumstances. A second reaction is to state that there is some quality of the individual committing the infraction that is not representative of the ingroup. A third reaction is to explain why the action is not as bad as it seems.

When someone is an outgroup member, though, this tendency reverses. When an outgroup member does something good, he is considered unusual—his actions do not reflect positively on the entire outgroup. But when he does something negative, it confirms that the outgroup really is as awful as you suspected.

In the National Football League play-offs following the 2014 season, quarterback Tom Brady and the New England Patriots coaching staff were accused of cheating by deflating the footballs to make them easier to grasp in cold weather. Fans of the Patriots (who consider Tom Brady to be an ingroup member) interpreted his actions as either necessary, given the cold weather, or else a sign of the team's competitiveness (which is a good thing, right?).

For fans of every other NFL team, though, Tom Brady's actions were a sign that he is just a typical member of the New England Patriots outgroup, who will do anything to win a football game, no matter how underhanded.

This tendency to interpret the actions of ingroup and outgroup members in a way that is consistent with previous beliefs about the groups makes it hard for people to change their opinions. After all, if you interpret every negative thing about an outgroup as a sign that the outgroup is bad, you reinforce your fear or distaste. And if you don't allow that broad belief to be affected by positive interactions with particular individuals, then the belief will be hard to change.

THIS DISTINCTION BETWEEN INGROUPS AND OUTGROUPS APPEARS EARLY. EVEN young children tend to prefer people who are like them over people who are not like them. Indeed, the minimal groups technique works with young children, who will prefer someone wearing the same color shirt or even someone who selects the same kind of food that they like.

One big difference between young children and adults, however, is that although young children have a significant preference for people who are like them, they do not have the same tendency to dislike members of outgroups. It seems that children are actually conditioned by society to believe that people who are not like them should be viewed with suspicion—whether those differences involve appearance, dress, or language.

The mistrust of strangers may have made a lot of sense early in our evolutionary history. In prehistoric times, neighboring tribes may have been hostile, particularly when resources were limited—at least until groups began to join forces and cooperate with one another, forming the basis of the large-scale societies we have today. And it was probably a good strategy for small bands of people to have mechanisms to bind them together. After all, if the social group is small, everyone needs to work extra hard to maintain harmony, to protect one another, and to feed and care for one another. Mechanisms that would make the group seem like an extension of the self would help groups work together to overcome adversity.

It is difficult to know for sure what made sense in our evolutionary history, of course, because there were no observers taking notes. As a result, evolutionary explanations like this can become "just-so stories" (named after the famous set of stories by Rudyard Kipling that gave fanciful explanations describing how various natural phenomena—like how the leopard got it spots—came to be).

The modern developed world does not resemble the environments of our evolutionary past. There are certainly occasional dangers, but societies have done an excellent job of making the world a generally safe place. In addition, the modern world permits an unprecedented degree of travel. Pile into a metal tube with wings, and you can be anywhere on the planet in a day (barring travel delays). The global economy allows people from many different cultural backgrounds to mix together. Hence, the diversity of the environment where Art was living in New York.

The solution to prejudice and dislike of strangers is diversity of interaction. Humans are very sensitive to the statistics of our environment. Even though we may have an initial mistrust of strangers, as those strangers become a commonplace part of our environment, learning mechanisms turn familiar people (and familiar types of people) into part of the neighborhood ingroup rather than the stranger outgroup.

SOCIAL INTERACTION TURNS OUTGROUPS INTO INGROUPS.

32

What's the best way to deal with life's endless litany of nuisances?

THERE ARE LOTS OF ANNOYING THINGS IN THE WORLD. YOU PROBABLY have acquaintances or colleagues whom you dread seeing, knowing that you are about to get dragged into another pointless conversation without end. Driving has its share of annoyances as well. Other drivers on the road are always driving too fast or too slow, or not signaling, or riding their brakes. And if you live in a place like Austin, there is more traffic on the road every day.

It can get maddening.

And that can lead you to do something you might regret later. Years ago, Art was pulling into a parking garage at work. He started to drive through the entrance at the same time as another car. The other driver honked and then waved with one finger. As their eyes met, the driver of the other car realized (to her horror) that she was flipping the bird at one of her colleagues. They are still able to laugh at that incident, but if they had it to do all over again, perhaps avoiding that little incident of road rage would have been a good idea.

Why do these kinds of annoyances bubble over into frustration and (sometimes) anger?

A big piece of this has to do with a concept called *locus of control*. At any given moment, your actions are governed by a combination of the decisions you make and the situation you are in. Let's leave aside the thorny question of whether there really is free will. We all certainly feel as though we have some *agency*—some ability to act on the world and affect it—at least.

When you feel that you are able to affect the outcomes of the world around you, then you are experiencing an *internal* locus of control. That is, you feel as though you are the author of your destiny. But when you feel that the world is dictating your fate, then you are experiencing an *external* locus of control. That is, you feel as though you have strapped yourself into a roller coaster, and there isn't much you can do to change what is happening.

The two of us both generally feel as though we are affecting the world around us more than the world is dictating out circumstances, so we generally experience an internal locus of control. But we both know plenty of people who feel as though there is little they can do to change what is going on around them. Those people walk around with an external locus of control.

In the extreme, an external locus of control can lead to what is called *learned helplessness*. We talked about learned helplessness earlier when we were discussing whether you can make yourself happy. As we mentioned, some people find themselves in circumstances in which they have no real options to affect the course of their lives. Those people may slowly give up hope that they can act as agents in the world. Eventually, they may just stop trying.

Learned helplessness understandably leads to bad outcomes. Think about a boy in elementary school who is suffering from dyslexia. Every other child in the class looks at books and can clearly identify words and sentences and obtain information, but no matter how hard the child with dyslexia tries, the books don't work their magic. If the learning disability goes undiagnosed, then this child will eventually feel that there is nothing he can do to be a better student. At that point, he might just give up in school. Even if the dyslexia is eventually diagnosed, he may no longer feel that he can affect the course of his education.

The same kind of thing can happen to people who are dealing with a chronic illness. After suffering for years, they may just decide that there is nothing they can do to stay on top of it. Without hope, these patients may no longer expend effort to fight their disease.

Of course, this is the extreme version of what happens for people with an external locus of control over a long period of time. In the short term, an external locus of control often leads to frustration and anger.

THINK ABOUT WHAT HAPPENS WHEN YOU ARE SITTING IN TRAFFIC. BOB IS normally a placid driver. In his cute little Smart car, he enjoys tooling around the streets of Austin. When he hits the inevitable traffic delays, he generally handles them with aplomb. But—every once in a while—he is running late for an appointment, and the traffic is unexpectedly heavy. And then a driver in front of him decides to change lanes without signaling and ties up traffic even more.

And that is when the frustration starts to build up.

In this case, Bob (who generally has an internal locus of control) is stuck in a situation in which there isn't much he can do. The world is making him late for an appointment. And there are boneheaded drivers on the road, to boot!

What should Bob do in this situation?

You might think that the answer is to give in to the frustration and to yell. Lean on the horn, perhaps. Get the anger out of his system. This reaction reflects a belief in *catharsis*. One way to think about this is that anger is like heated water in a boiler. As the water heats, the pressure builds up. The only way to keep the boiler from exploding is to let off a little steam. The yelling and aggression are attempts to decrease the pressure. Sometimes when you yell and scream and make rude gestures, you do feel somewhat better.

Unfortunately, catharsis doesn't work so well in the long run. It turns out that your brain is a habit-creation machine, and it will try to associate the environment with the behavior you find yourself performing most often in a given situation. If you yell and scream whenever you start to feel frustration, your brain starts to assume that this is the proper response to that emotion. You learn that frustration is a signal to act aggressively. And that can cause problems.

Luckily, Bob takes a different approach to such frustrating situations. First, it is helpful to recognize that, at this moment, the frustrating circumstances are out of your control. Yelling at other drivers is not going to change anything. Instead, it is useful to find ways to distance yourself from the immediacy of the situation. Bob will think about the fact that the worst-case scenario is that he misses the appointment and has to reschedule. He will turn on the radio and listen to some soothing music. He will breathe a little more deeply, because deep breathing is relaxing.

When there is nothing you can do about a situation, don't give in to the frustration. When you distance yourself from it mentally, the situation has less impact on your emotional state. And when you engage in relaxing activities like deep breathing, you pull more of the energy out of the situation, which makes you feel more in control and prevents feelings of helplessness and rage, which take a toll on your mind and body. Yes, you have no control over your arrival time at the appointment and you might have to reschedule it, but you're not going to wreck your mental and physical health in the process.

OF COURSE, THERE ARE FRUSTRATING MOMENTS WHEN IT SEEMS LIKE THERE IS nothing you can do to change the situation, but actually you can. Suppose you have an Eeyore in your life—a person who finds the cloud in every silver lining. You may dread conversations with this friend because he invariably descends into a gripe-fest. You might even feel the frustration building when you see this person walking up to you unexpectedly. You brace yourself for another unpleasant interaction.

Before this individual even starts talking, control the direction of the conversation. Find something upbeat and positive to say. Kill the negative energy with kindness. One of two things will happen: Either the conversation will turn out to be a pleasant one or the person will seek out someone else who is more interested in being negative. Either way, you have reframed the situation by taking back some control. It turns out that there are lots of situations in which you may feel like you don't have any control, when actually you do.

When life gives you LEMONS, make a fabulous LEMON DROP MARTINI (but not while driving).

33

Is mind reading a necessary skill?

ART HAS TWO DOGS. THEY SEEM TO LIKE HIM WELL ENOUGH, BUT THEY don't understand him very well. While he is standing in the kitchen, they will put their paws up on the counter looking for scraps of food, even though they have been repeatedly moved away from their prize before they can get a bite of whatever is being prepared. They just don't seem to understand that if Art can see them, they aren't going to get away with stealing food. On some occasions, the dogs stumble into the kitchen and hit the jackpot, walking away with a choice morsel. But even then, the dogs don't seem to know that it might not be the best idea to walk back into a room where Art is sitting while continuing to chew on what they have stolen.

Kids are much smarter than that. The two of us have both raised kids who learned early on that if they wanted to get away with something that their parents didn't want them to do, like sneak a cookie before dinner, it was best to do it out of sight. They will wait until the coast is clear and then go for the spoils, making sure to enjoy their bounty away from their parents' watchful eyes. Actually, getting away with a successful cookie-theft requires some sophisticated reasoning. Kids need to understand what other people (namely their parents) know. Kids need to recognize that they are trying to create a difference in beliefs between themselves and their parents. That means that they need to know about all the ways that parents can find out about cookie-eating, including (1) seeing a cookie being taken and eaten, (2) evidence that cookies are missing, and (3) evidence of illicit cookie-eating, such as crumbs in the bedroom.

This ability of kids (but not dogs) to get away with cookie-theft reflects the fact that people have a pretty good *theory of mind*. That is, they are able to think about what they know versus what other people know and to keep those things separate. This requires an understanding of the factors that allow another person to know something.

Let's pick this apart a bit.

People understand all the ways that other people can get information. If people see an event happen, then they will probably know about it. If they read about the event, they will know about it. If they hear about it from someone else, they will know about it. The more evidence of an event is lying around, the more likely it is that people will find out about it. But in the absence of any contact with information about the event, they will not know about it.

This set of beliefs about when someone is able to learn a piece of information is helpful because it allows you to get away with all kinds of stuff without tipping off other people. If you know that not everyone knows what you know and that there are things that other people know that you do not, you can understand why people hold different beliefs. And because people differ in their beliefs, you can understand why people believe things that are false.

For example, suppose John walks into the kitchen and sees Mary putting cookies into the cupboard next to the refrigerator.

Mary doesn't want John to know where the cookies are, so after John leaves the kitchen, Mary takes them out of the cupboard next to the kitchen and puts them in a drawer by the stove. If I ask you where the cookies are now, you will know that they are in the drawer by the stove. If I ask you where John *thinks* the cookies are, though, you will recognize that he still thinks they are in the cupboard by the refrigerator.

That Mary and John hold different beliefs about the location of the cookies might seem obvious and not terribly interesting, but it actually takes a while for kids to learn to keep all of this straight. Young kids are often tested in *false belief tasks* in which a character sees an object in one location and then leaves the room while a second person moves the object to a new location. When children are two or three years old, they have a hard time keeping straight that the person who left the room does not know where the object has been moved to, even though the child knows. Very young children might not have enough working memory capacity to keep track of who knows what, but by the time they are five or six years old, they do a pretty good job on these tests.

Because it takes some mental effort to keep track of who knows what, even adults don't always do it well unless they are motivated to do so. In a clever study done by Boaz Keysar, adults were told a story about Michael, whose parents were coming to visit. He asked his secretary for a recommendation for a restaurant. She suggested a new Italian place, so Michael took his parents there for dinner. As it turned out, the meal was absolutely horrible. In the morning, he left his secretary a quick note that said, "Thanks for the recommendation—the meal was great, just *great*."

After reading this story, people were asked how the secretary would interpret the message. Many of them said that she would interpret the note sarcastically. Now, the only way she could think her boss was being sarcastic is if she knew that he'd had a terrible meal, but she has no way of knowing that. (Of course, if she worked for Bob, she might just assume he was being sarcastic based on past experience, but the readers had no way of knowing that, either.) Instead, it seems that readers just weren't that motivated to keep track of which characters had which pieces of information in the story, so they didn't bother to create separate files to remember who had which pieces of information.

So what good is knowing what other people know?

The examples we used to start this chapter show that having a theory of mind is helpful if you want to deceive others. Dogs are not particularly good liars because they don't seem to realize how people come to learn new information, nor do they really reason about who knows what. That is not to say that dogs aren't clever. One of Art's dogs is much more likely to jump on the counter when Art leaves a room than when he is there. That is not because she is able to reason about Art's mind. Her *dognition* is probably not that sophisticated. Instead, she has just learned that she gets rewarded with a taste of food more often when he is not there than when he is.

A theory of mind isn't just for lying, though. Actually, most successful communication requires that each person understand both the knowledge they are likely to share as well as the knowledge each speaker has individually. Suppose Art and Bob are having a conversation. They can draw on a lot of *common ground* for their conversation because they are friends and because they are both psychologists. They can refer to shared events that they attended. They can talk about mutual friends. They can also talk about professional knowledge that they believe most other psychologists would know.

If Bob tries to tell Art something that they both know that Art already knows, however, the interaction is going to be frustrating.

Generally speaking, you are trying to tell people things that they do *not* already know. Why would you tell someone something if she already knows it?

When writing this book, we had to make a guess as to what our readers are going to know. Some concepts are ones that we assume are part of the shared knowledge that almost every reader will have. We started this chapter talking about dogs with the assumption that readers will know what dogs are and that they are often kept as pets (even if they didn't know that Art had dogs). We generally assume that any psychology concept is going to be a new one for readers, though, so we spend some time explaining those concepts in detail.

Without a reasonable sense of what's in other people's minds, it's difficult to interact effectively. In fact, a number of researchers have speculated that one characteristic people with autism share is a faulty theory of mind. You can imagine how debilitating it would be, from a social perspective, if you could not really understand what other people were thinking, what they knew, and what they did not know.

On the flip side, having a theory of mind makes lying possible. Studies with preschoolers have demonstrated that if you teach children how to distinguish between their own beliefs and the beliefs of other people, they immediately start to use that knowledge strategically by lying.

By allowing us to understand what other people know and what they don't, theory of mind is crucial to the ability to communicate and interact effectively with others. Yes, an advanced theory of mind may give rise to the most skilled artists of deception, but any good tool can be misused.

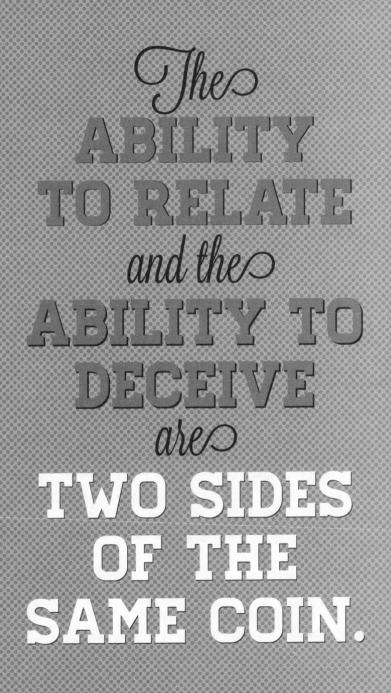

The
ABILITY
TO RELATE
and the
ABILITY TO
DECEIVE
are
TWO SIDES
OF THE
SAME COIN.

34

What are brains for, anyway?

THE BRAIN SEEMS SO IMPORTANT FOR LIFE THAT PEOPLE RARELY QUEStion what the brain is actually for. It seems like an odd question. Why do we even need to ask?

Even though it seems way better to have a brain than not to, it's biologically expensive to make a brain, and it takes a lot of energy to run one. Human brains typically represent about 3 percent of body weight, but they consume around 20 percent of the body's energy, all day every day, even during sleep. Most species don't have big brains, and most living things on the planet (like single-celled creatures, plants, and fungi) have no brain at all.

So how did the selective pressures of evolution determine that all that energy is worth it? Everyone who's seen the series of silhouettes depicting fish turning into reptiles, turning into rodents, turning into monkeys, turning into us might assume that evolution is a progression toward greater and greater complexity. But that's not actually how it works. The course of evolution isn't a line or even a tree. It's a dense bush with lots and lots of branches and twigs that don't ever lead to anything. Most of the gazillions of genetic mutations that form the basis of evolution either don't have any effect at all or lead to changes in organisms that ultimately don't survive. Every once in a while, though, a genetic mutation produces a change that conveys a survival advantage, and when an organism survives, it can pass on the advantageous mutation to future generations.

Our earliest single-celled ancestors had nothing at all resembling a brain. Their behavior was governed by chemical reactions to whatever was going on in their environment. When single-celled organisms were tightly packed together in structures like biofilms, and when individual cells serendipitously combined to form multicellular organisms, chemical communication—dispersing molecules that other cells could detect—sufficed. But one day, one lucky species developed a way to signal other cells with electricity.

Chemical signals diffuse slowly and haphazardly through water (which is the dominant medium of cells and the environment around many early organisms). Electrical signals, on the other hand, can travel faster and farther than chemical signals, so the ability to send these signals conferred an advantage to that lucky "electric" ancestor. Over countless generations of species these cells eventually formed into clumps called *ganglia*, and those clumps of cells that signaled using electricity (nerve cells) developed the capacity not only to pass information but also to change as a result of experience: to remember what happened—that is, to *learn*—by turning on and off the gates that allow ions into and out of the cell. Thus began the evolution of the three pounds of goo we're all lucky enough to carry around inside our skulls.

WHY WAS THE ABILITY TO LEARN SO IMPORTANT? AS SOON AS A BRAIN CAN store information about the past, it can allow the organism to change its reactions to future situations based on experience. The organism can *predict* what the future will hold. And that provides a survival advantage over creatures that cannot make predictions.

If you're wandering around in the dark, not knowing where you are, what's around you, or what whatever-it-is that's around you is going to do, you're pretty vulnerable. Think about that for a moment. Predators might catch you unawares (and you're dead), you might miss opportunities to obtain food (dead), or your competitors might get all the good mates (alas). You wind up at the end of one of those short twigs on the evolutionary bush. But if you can predict what's about to happen, the world is a little more navigable. You're better able to catch escaping prey if you can anticipate how they turn when they run. You're better able to escape a predator if you can gauge the size of hole that you can fit into but the predator can't. Brains are able to carry out this ability to predict and respond accordingly on a grand scale.

As you move around in the world, your brain constantly makes predictions based on what's already stored in your memory. Most of the time those predictions turn out to be pretty good. You can grasp a cup, turn the key in the ignition of your car, smile at a friend, or extend your arm to shake the hand of a new acquaintance, and your brain predicts what comes next. What *actually happens* in each of those experiences has the potential to update your memory in ways that make your future predictions more reliable. If you're cooking with a new frying pan, and the first time you grab its metal handle you realize that it conducts more heat than the handle of your old pan, the hot handle and the brief pain it produces update your memory about holding a frying pan. Each little prediction error and correction creates important changes in the brain that make survival more likely. Ultimately, the expensive brain is worth having; it makes your life more manageable and your efforts more successful than those of organisms without a brain, left to respond to the vicissitudes of life having no idea what might happen next.

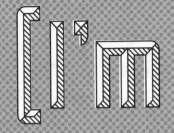

Does listening to Mozart make us smarter?

WOULDN'T IT BE GREAT IF GETTING SMARTER REQUIRED NO EFFORT AT all? What if you could just crank up *Eine Kleine Nachtmusik* and watch your IQ tick up with every passing note. Yes, it would be great if listening to music did that. But as the saying goes: If it sounds too good to be true, it probably is. And so it goes for music and intelligence.

So how did the claim that "Mozart makes you smarter" become accepted wisdom if that's not really what happens? Well, it started with the journal *Nature*, one of the two or three most prestigious scientific publications in the world. In the early 1990s several psychologists published the results of a study in which a small group of the most ubiquitous subjects of psychological research, college undergraduates, took a test of a particular dimension of thinking called *spatial reasoning*. A third of the students listened to ten minutes of Mozart before they took the test; another third listened to a "relaxation tape"; and the remaining third sat in silence for the same amount of time. Ta da! The students who listened to the relaxation tape or sat in silence scored about the same, but the Mozart listeners scored slightly higher than their peers on the spatial reasoning test. The authors of the study proposed that somehow the structure of Mozart's music activated responses in the brain that led to the higher scores. Cue the media.

Seemingly overnight, the Mozart Effect® was everywhere. The governor of Georgia devoted $100,000 of the state's budget in 1998 to provide every new Georgia mother with a recording of Mozart's music to play for their babies when they left the maternity ward. New businesses sprang to life, claiming that a little Mozart does a lot of good. And even now you can purchase CDs called *Mozart for Brain Power*, *Mental Mozart*, *Mozart for Your Mind: Boost Your Brain Power*, *The Mozart Effect: Music for Newborns—A Bright Beginning*. We could go on, but you get the idea.

Of course, in the experiment, listening to Mozart was compared with relaxation and sitting quietly, so there were lots of differences between the conditions: silence or relaxation versus . . . anything at all—sound, music, piano music, classical piano music, classical piano music by Mozart, a particular piano sonata by Mozart (Sonata for Two Pianos in D Major, K. 488). The authors preferred the last, most specific interpretation.

Soon after this, though, other psychologists began to test the supposed effect in their own labs. Some found that listening to many different kinds of music, even music by Yanni, produced similar effects. We shudder to think of a boom in the Yanni Effect (yikes). One research group produced similar results by having students stare at a colorful computer screen saver before taking the test. And in one of the most dramatic claims about the Mozart Effect in nonhumans, rats that were raised listening to Mozart (even during gestation) ran mazes faster than less fortunate rats that listened to white noise or the minimalist music of Philip Glass.

It was later pointed out, somewhat embarrassingly, that (1) rats can't hear in utero, (2) rats are born deaf, and (3) the rat auditory system is unable to perceive most of the notes of the Mozart piece they were being exposed to. Hmm.

So what's actually going on with all of this? The prevailing interpretation is that arousal and enjoyment are actually responsible for better mental performance. In a number of experiments testing the supposed Mozart Effect, the more the participants enjoyed the stimulus—whether the stimulus was music, a good story, or nature sounds—the more positively it affected their scores. And in fact, there is a lot of good evidence out there that a positive mood makes you more creative and allows you to perform better on a variety of tests.

WE DECIDED TO WRITE ABOUT THIS NOT ONLY BECAUSE OF WHAT IT TEACHES us about the potential of music listening "to make us smarter," but also because of what it teaches us about the process of science. Of course, all of us who conduct experiments develop a fondness for our explanations of the world, and that fondness sometimes persists even in the face of irrefutable evidence that we're mistaken. We're human beings, after all—prone to all of the irrationality that comes with being human. This was perhaps especially true in the case of the Mozart Effect, because much of the media, not to mention the National Association of Music Merchants®, had invested a great deal in touting the effects of music listening on intelligence. And as Upton Sinclair famously stated (pardon the gender reference), "It is difficult to get a man to understand something when his salary depends on his not understanding it."

The Mozart Effect craze is emblematic of what happens when ideas that allow you to get a positive outcome without effort gain a patina of scientific credibility. I can get smarter, and I can make my children smarter, just by listening to a certain kind of music? What's not to like about that? And there was an article published about it in *Nature*, for goodness' sake. (The same kind of craze develops whenever we find out that we can eat chocolate and lose weight, or drink red wine and live forever.)

But science was designed to prevent us human beings from engaging in wishful thinking. Good science defines the rules of evidence and clarifies what counts as evidence *before* we run an experiment, and that very powerful set of rules has led to our species' figuring out all of what we understand about the world that is not intuitively obvious. From our human scale and perspective, there is no indication that we are standing on the surface of a rotating sphere that is hurtling through space at sixty-seven thousand miles per hour. It looks pretty peaceful out there in the backyard. We see no evidence of the fact that over half of the cells in and on our bodies are not human cells, but bacteria and other microbes. Eeeek.

Science cynics might contend that it was science—published in one of the most prestigious journals in the world!—that started the whole Mozart craze. True enough. But it was also science—which insistently requires the replication of experiments to verify their credibility—that eventually corrected what was an initial misinterpretation of how the world works.

BEFORE WE LEAVE THIS STORY, IT'S IMPORTANT FOR US TO POINT OUT THAT learning to *make* music is another kettle of fish entirely. There are multiple benefits of learning to sing and play an instrument. We know that music making engages more of the human brain than just about any other activity. Making music involves your perceptual system, your motor system, your emotional system, and your motivational system in a coordinated activity that has the added advantage of being tremendously satisfying.

Art had told himself for years that he wanted to learn to play the saxophone, and rather than waiting till the end of his days, only to whisper to his loved ones gathered around his bedside that "I always wanted to play the saxophone," he actually went out and *learned to play the saxophone*. In fact, he still takes lessons and plays in two different bands.

He didn't learn an instrument to improve his thinking (though it couldn't hurt); he learned because he loves music and wanted to be an active participant in its creation. Interestingly enough, playing music does make you smarter, if what you mean by smarter is better at coordinating the many aspects of thinking, behaving, and feeling that are involved in making music. And even if it doesn't make you smarter, after you learn to play an instrument, *you can play an instrument!* That is cool. And fun.

Ideas like the Mozart Effect, with their accompanying CDs and other paraphernalia, are filed away in the folder called "Magical Thinking." As we've written about in other chapters, improving your brain requires effort and attention. Listening to music is a joy. Making music is a different kind of joy that has the added benefit of developing your brain.

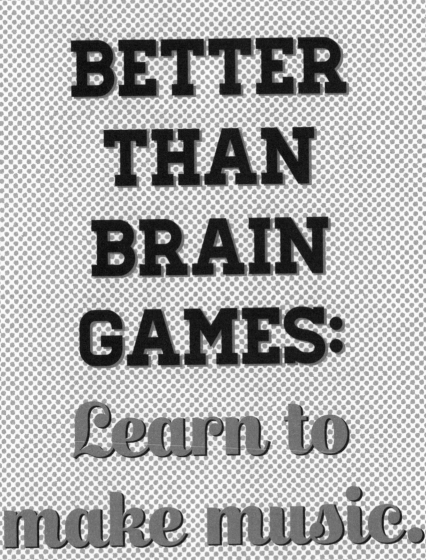

BETTER THAN THAN BRAIN GAMES: Learn to make music.

Why are
other people
such slackers?

THE TWO OF US HANG AROUND A LOT OF MUSICIANS, AND OVER THE years we have both been in bands. It is amazing how a great band can disintegrate. One moment, a group of musicians are working together making incredible music. The next, they are throwing things at each other, having horrible arguments, and wishing they had never met.

There are lots of forces that can lead a band to split up. It is never easy living with a group of other people on tour, for example. Those long hours on the road can bring out the worst in people. It is hard enough living with people when they are part of your family and everyone has a private space.

But the biggest killer of bands is the *credit assignment problem*.

Everyone in the band (plus the manager, album producers, and others) plays some role in the band's success.

Suppose the band had a psychologist (or perhaps two) who got each person associated with the band to rate what percentage of the success of the band was due to their efforts independent of everyone else's. What would happen?

If everyone in the band were pretty well calibrated about their contributions to the band's success, then everyone's ratings would add up to 100 percent. If the band's members were focused mostly on the contributions everyone else made, they might even underestimate their own influence on the band's success, and the ratings would add up to less than 100 percent.

In actuality, just about everyone overestimates their contributions, and the ratings add up to way, way, way more than 100 percent.

The idea is that (just about) everyone has an *egocentric bias*. That is, we are much more focused on our own actions than everyone else's, and that causes us to highly overestimate our contributions to a group effort.

Where does this egocentric bias come from? There are several things that come together to lead people to focus extensively on their own efforts rather than on the work that other people have done.

Part of what is happening is just a function of memory. If you are a member of a band, you are able to recall many of the things you did that helped the band succeed: the times that you stayed late to clean up a studio after everyone else left; the hours you spent perfecting a riff on a new song; the moments of inspiration that led you to write a new tune that the band adopted.

You have witnessed other people's efforts as well, of course, but you haven't actually seen all of them. Much of their effort is hidden from you. So when it comes time to rate effort and you start to think about everyone's work, it's no surprise that *your* work is more salient to you than other people's.

In addition, *construal level effects* kick in. The more distant you are from something, the more abstractly you think about it. You are maximally close to yourself, of course, so you think about your own actions quite specifically. You think about the minute details of the things you do, but you think more abstractly about what other people have done.

If you write a song, you focus on specific actions: messing around with a particular riff, perfecting a chord sequence, adding a bridge, working through lyrics. These specific actions are effortful. When a bandmate writes a song, you think about the process without focusing on all the specific actions that were a part of making it happen.

Motivation also affects the way you value your contributions to a group. In particular, you tend to like things that are your own better than things that belong to someone else. This phenomenon—called the *endowment effect*—is true for objects and also for actions.

As an example, several years ago Art moved to a new house. Before moving, he held a garage sale—that time-honored method for unloading the junk that accumulated in the house. In preparation for the garage sale, he laid out the "merchandise" in the garage and driveway and started putting prices on everything. He was hoping to get $25 for a nice suit he had. Old tapes and CDs should fetch 50 cents to $1 apiece. A barely used bread machine was worth at least ten bucks.

Garage sale fanatics began to descend on the house. No matter what price Art had put on items, people asked to pay much less. Now, Art had expected some amount of bargaining, but he was surprised to discover that if he held the line on the price he marked, people just walked away. He figured that the people who came early only wanted big bargains, so he waited for the next round of folks to arrive.

There was a steady stream of people throughout the day, but they didn't seem that interested in paying the prices Art had put on the items. Eventually, he was selling things for almost no money. A couple of dollars for the blender. The suit went for $5, with a few shirts and ties thrown in.

It isn't just that people who go to garage sales are only looking for bargains. You value your own objects more than other people value them. Those objects have a history. And—more important—they are *your* objects. You instantly create an attachment to things that are yours, and that increases your sense of their value.

The same thing happens for your actions. Going back to the band, you have a particular love of the songs that you contributed to. You give extra weight to the actions you take on behalf of the band. So not only do you remember a lot more of your work than your bandmates', but you value your own work more than theirs as well.

ALL OF THIS BECOMES A PROBLEM WHEN PEOPLE ARE TRYING TO GET CREDIT and recognition for what they have done. This credit isn't necessarily financial (though for some bands it may be). Instead, people want others to acknowledge the efforts they have made.

It turns out that you can smooth out a lot of your social interactions if you spend some time acknowledging the efforts of others. It is useful to remember that all of us want to contribute as much as we can to the partnerships and organizations we belong to. Just as you are doing your best to make sure the group functions well, others are, too.

When you begin to feel as though you are doing more than your fair share of work, start by trying to find evidence of what others have done, and acknowledge their contributions. Expressing gratitude like this for what other people have done is a great way of reminding yourself of the efforts other people have made.

Bob is great at counteracting the effects of egocentric bias. He is always thanking people for work they have done and reminding people of what group members have done. That is an important part of why he is so happy much of the time. By thanking others, he is actually reminding himself of how much other people do for him. And that feels good.

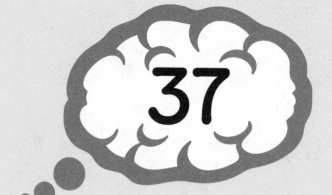

Can delusion be a good thing?

ELUSION IS JUST NOT A POSITIVE WORD. ART CAN STILL HEAR CLASSmates from high school yelling at one another, "Are you *deluded?*" to signal some off-base comment. When people are deluded, they believe something that is deeply at odds with reality.

Now, this book should make you feel less than entirely confident that the human brain always makes accurate assessments of the world. Because you reconstruct memories of the past, you can't be certain that a particular memory actually reflects the exact details of a previous event. Your beliefs about the world are not always coherent, either, so there are probably many contradictions among pieces of information stored in your long-term memory.

That said, while the visual system can sometimes be fooled by a particular view of an object, it uses so many different pieces of information to reconstruct the world around you that in natural situations you are rarely completely mistaken about what you see. A good artist may create a convincing illusion, but nature does not.

And that makes sense. It would be dangerous if you were often mistaken about what you see. You might fail to run from a dangerous situation. You might walk onto a surface that is not actually there. You might have difficulty locating and engaging with objects in the world if you did not have good information about what is around you. So evolution has placed a premium on maximizing speed and accuracy of vision. As a result, you see the world more or less the way it is.

There would seem to be an analogy between vision and the ability to conceptualize the world. Suppose you thought you were capable of lifting an object that you really cannot. You might waste a lot of time and effort trying to pick something up that won't budge. If you overestimated your skills at work, you might take on assignments that you cannot handle, which would get you fired.

In fact, there is a lot of evidence that people are systematically biased in their perception of reality. That is, many people are subject to delusions.

One great example of a common delusion is called the *Lake Wobegon Effect*, named after the fictional town in Garrison Keillor's long-running radio show on NPR. He describes Lake Wobegon as the town "where all the women are strong, all the men are good looking, and all the children are above average."

The Lake Wobegon Effect is the observation that people overestimate their skills in most tasks compared with the population at large. That is, if you ask people how good they are at a particular skill compared with other people and then you average together people's responses, you find that this value is above 50 percent. In other words: On average, people think they are above average.

The Lake Wobegon Effect is related to the findings of David Dunning and Justin Kruger described earlier, in the chapter on failure. Their research suggests that the worst performers on a task are the least well calibrated about their abilities. Poor performers know they are bad at something, but they don't realize quite how bad they are. Great performers, on the other hand, are often quite well calibrated about their performance because they need to focus on their errors in order to improve.

The big question is: Why does the Lake Wobegon Effect exist? Is there any advantage to thinking that you are better at a task than you really are?

To understand the potential advantage of the Lake Wobegon Effect, it is important to consider one other miscalibration people have: overconfidence about good outcomes. When entrepreneurs start a business, for example, they overestimate their chances of success. Students often overestimate the likelihood that they will get a good grade on a test.

This kind of overconfidence as well as the Lake Wobegon Effect are partly a result of motivated reasoning. That is, there is a general tendency for people to see the world as they want to see it. Believing that things will work out in the long run reduces anxiety and makes you feel better about life, so people bias their beliefs in the direction of the desired outcome.

However, the most significant relationship between miscalibrated beliefs and motivation actually goes in the other direction: People use their beliefs to help motivate them to succeed.

Theories of motivation point out that in order to really engage with some tasks, you have to believe that the goal you want to achieve is valuable. You also have to believe that hard work is required to achieve your goal, but that the goal is achievable if you put in the work.

If someone told you that he would pay you $1,000 if you just breathe for the next thirty seconds, you might be excited about the prospect of getting the money, but you probably wouldn't work too hard to breathe, since that is what you do naturally. If someone approached you on the streets of New York City and offered you $1,000 if you leapt from the sidewalk to the top of a skyscraper, on the other hand, you wouldn't even bother to try because the task is clearly impossible.

In the middle range between the natural and the impossible, though, you *do* work hard. And it pays to be a bit delusional. For example, Bob went to college to study music performance. It is hard to make a living as a musician—most of the musicians we know also work other jobs to make ends meet—but you have to believe that your situation will be different from everyone else's if you make the decision to pursue a degree in music. To that end, it helps to be overly confident that you will succeed.

Bob eventually decided to go back to school to become a teacher and finally to get a PhD, do research, and teach at the college level. This choice also required some overconfidence. Faculty jobs at universities are scarce, and far more people get PhDs each year than there are available jobs at universities. If potential PhD students who aspire to teach at universities were realistic about their chances of achieving their goal, they might never start on the road toward a PhD.

So people bias their perceptions of the world in ways that create motivation to work toward a goal. They overestimate their chances of success in order to make the goal seem achievable. They overestimate their own abilities in order to enhance their belief that they have the skills to succeed if they try. These are examples of delusion put to good use.

In fact, this combination of miscalibrated beliefs can lead to a self-fulfilling prophecy if you work hard. Working hard toward a goal allows you to improve your skills and increases the chances that you will actually achieve your goal. That means you may be more likely to succeed by having poorly calibrated beliefs than you would be if you were precisely accurate in your assessments of yourself and the world—which could very well stop you in your tracks.

OVERCONFIDENCE IS NOT THE ONLY WAY THAT YOU CAN USE YOUR BELIEFS TO manage your motivational system. You have probably engaged in *defensive pessimism* as well. That is, you assume that you are going to perform poorly on an exam, at a task at work, or in a presentation.

When you engage in defensive pessimism, you generally assume you are going to perform even more poorly than you actually do. The belief that you will do poorly creates anxiety. Your motivational system is engaged to reduce that anxiety, which leads you to work harder to prepare. That hard work generally improves your performance.

For optimal results, the best strategy is to find the ideal gap between where you are right now and where you hope to be. If that gap is too small, then you need to work to broaden it. Defensive pessimism makes you think you are further from the goal than you actually are, and that increases motivation—provided you don't abandon the goal and resign yourself to failure.

If, however, the gap between where you are and where you aspire to be is too large, overconfidence in your abilities and in your likelihood of success can be used to narrow your perception of the distance you need to travel.

To succeed, MIND THE GAP (between yourself and your goal).

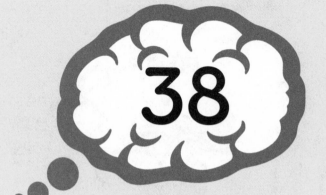

Why do we call
a dog a "dog"?

YOU STEP OUT ON THE STREET AND LOOK AROUND. IF ASKED TO DE-
scribe what you see, chances are you will point out several
cars, some people, a dog or two, some bicycles, houses, and
stores. None of this seems at all surprising.

But think for a moment about the language you are using to
describe this scene. You identify a particular object as a "car." Why
did you use the word *car*? You could have said that it was a "vehicle," or perhaps a "2012 Ford Focus®." What made you choose words
like *car*, *house*, and *bicycle* rather than some of the alternatives you
could have selected?

Whenever you categorize things, you can select categories that
group objects at different levels of specificity. When you call something an "animal," you are lumping it in with a large number of the
creatures on planet Earth. When you call something a "dog," you
have identified a particular group of four-legged furry barking creatures. Still more specifically, when you call something a "poodle"
or even a "standard poodle," you are narrowing the category down

even further. And when you talk about "Guido, the standard poodle who lives next door," you have gotten this all the way down to a particular (rather cute) dog.

Psychologists have recognized for a long time that the middle level of specificity is the one that is most commonly used when talking about and thinking about objects. In recognition of this importance, this middle level of specificity is called the basic level. So *dog* and *car* are basic level categories. More general categories (like *animal* and *vehicle*) are called *superordinate* categories. More specific categories (like *collie* and *Chevrolet Camaro®*) are *subordinate* categories.

Lots of studies demonstrate that basic level categories have a behavioral advantage over other categories. Children tend to learn basic level names for things before other labels. Adults usually use the basic level word to name an object. Basic level names are usually shorter than other words for objects. Many superordinate and subordinate category labels involve several words, like *construction equipment* and *standard poodle*.

Because of the importance of basic level categories, studies have also explored why that level has become the one used most often. There are two factors at work here. One is informational. Speaking and listening both require effort. The speaker has to choose what words to use to describe things in the world. The listener then needs to take the words the speaker is using and figure out what those words refer to. It turns out that it is not possible to minimize the effort of both speakers and listeners at the same time.

For speakers, it would be easiest if every item in the world could be given the same label. For example, suppose every object was called *thing*. Speakers could say, "Put the thing on the thing by the thing." Simple.

Unfortunately, sentences that use only general words like this make the task of decoding what the sentence means nearly impossible. To make it easiest for listeners, speakers would use very specific language. They might say, "Put the silver bread knife on the mahogany dining room table by the cut-glass decanter." This sentence would help to minimize the effort that listeners need to make, but speakers would have to work hard to create each sentence.

The basic level provides a compromise that requires both speakers and listeners to engage in a moderate amount of effort when conversing. Studies show that the basic level is well suited to help minimize that effort. This is just one example of many situations in which the brain tries to minimize joint effort.

ELEANOR ROSCH AND HER COLLEAGUES FOUND A NUMBER OF INTERESTING properties of objects that have basic level labels. Basic level categories are the most abstract, in that most of the objects have the same shape. Thus, all dogs look relatively similar, but their shapes differ from contrasting categories like cats, goats, and sheep. Screwdrivers look relatively similar, but differ from contrasting categories like hammers and saws.

Basic level categories are also the most abstract category for which most of the objects have similar parts. Chairs tend to have legs, backs, seats, and arms. Other types of furniture, like tables, have some similar properties (like legs), but not others (they don't have seats or arms). Similarly, dogs have legs, snouts, ears, and tails. Animals from other basic level categories (like cats, fish, and birds) share some properties, but not all of them.

For categories that describe human-made objects, basic level categories are also the most abstract level at which the objects share a function. Screwdrivers are used to install screws, though different types of screwdrivers are used with different types of screws. Saws are used to cut materials, though different types of saws cut different kinds of material or cut with different degrees of accuracy. Hammers are used to pound fasteners into materials, but different types of hammers are used for different kinds of fasteners.

For categories of living things, basic level categories tend to have similar behavioral properties. For example, dogs bark, cats meow, and cows moo—no matter what the breed.

BECAUSE THE BASIC LEVEL RESULTS FROM THE AMOUNT OF EFFORT THAT speakers and listeners need to invest in order to converse, there are several factors that push around what label people will actually use in a given instance. For example, a more general label may be used when it's obvious what a speaker is talking about. If there is only one object on a table, a speaker can say, "Hand me the thing on the table," because the listener will know what the speaker means. If there are many different objects from the same category, the speaker will often select a more specific label. At an animal shelter, it is not enough to say, "Look at the dog." Instead, a breed name would help, or perhaps other words describing the color and size of the dog.

Experts will often use more specific categories, particularly when speaking to other experts. For dog experts, for example, the breed of the dog is an obvious property, so it is easy to use this more specific category. Experts can also identify members of specific categories easily, so the specific category becomes the best one for them to use with one another.

Because we all expect to hear basic level categories in conversation, we can also use other levels of categorization to communicate things that go beyond the category label itself. That is, we can use the contrast between the label someone expects us to use and the label we actually use to communicate or emphasize something interesting.

Art is a dog lover, but his mother is really afraid of dogs. If Art walks into a room with one of his dogs, his mother might say, "Get that *animal* out of here." Ordinarily, you would expect someone to use the word *dog* in this situation. By saying "animal," though, Art's mom is highlighting the beastly, primitive aspect of the word *animal* as opposed to the loyal, friendly image implied by the word *dog*. Thus, she has played with category labels to clearly communicate her distaste toward Art's dogs.

Similarly, using more specific labels than is required can also communicate something about the speaker. Bob sometimes hangs around with snooty wine connoisseurs. A group of typical people sitting around a table might ask someone to pour more wine (and perhaps specify red or white, if there is an option). But a wine snob might be sure to use a more specific label like, ahem, the "2012 Côtes du Rhône." By using the more specific label, the wine snob is trying to convey a high level of knowledge about wines, in addition to just asking for more.

You can see by now why it is valuable to have many different labels for objects. It is useful to be able to group things at different levels of specificity. It is also helpful to have words that name the functions of things. For example, many dogs are also *pets*—a word that highlights the relationship dogs have to their owners. The more different ways there are to talk about things, though, the more effort it takes to communicate effectively. That is why it is good to have a straightforward strategy to select a label to use when talking.

And that's why we call a dog a "dog."

THERE'S MORE THAN ONE WAY TO SKIN A ~~MAMMAL~~ ~~FELINE~~ ~~SIAMESE~~ cat.

Why do we love kitten videos so much?

THERE ARE A SURPRISING (OR PERHAPS FRIGHTENING) NUMBER OF VIDeos on the Internet that involve small animals (particularly dogs and cats) doing adorable things. In the name of research, Art spent an hour on YouTube recently. In that time, he came across videos of kittens resting, kittens playing with dogs, puppies crawling along with babies, kittens chasing dots from laser pointers, cats trying (and failing) to jump on couches, and puppies playing with squeaky balls. Bob had to haul Art away from the computer after an hour to prevent a cuteness overdose.

What's with our seemingly boundless fascination with kitten videos?

The starting point for the joy of kitten videos is our evolutionarily programmed sense of what is cute. Both men and women (unless they have hearts of stone) find human infants cute. Particularly cute infants tend to have larger heads but smaller features. This structure gives infant faces large foreheads. Cute infants also have large eyes.

Infant cuteness creates positive feelings in the adults and older children who see them. This positive feeling is necessary because (as anyone who has spent a lot of time with infants can attest) babies can be quite difficult to be around in large doses. Babies require a lot of care. They need to be fed, washed, and carried. They often cry when they are not sleeping. And babies often wake up in the middle of the night needing to be fed, or changed, or just because.

In order for all of the baby demands to seem worth it to adult caregivers, they need some positive feedback. Before infants are mature enough to engage with adults and smile and laugh, all they have going for them is their cuteness. And not surprisingly, adults are wired to respond to infant facial features with positive feelings.

As it turns out, the facial features of immature humans are similar to the facial features of young animals, like kittens and puppies. Baby animals also have small facial features and large eyes, so you generally respond to kittens and puppies in the same way you do to babies. Looking at them, you get an immediate dose of the "warm fuzzies."

Kittens and puppies have additional features that make them fun to watch. They are not particularly well coordinated. They are prone to stumble and fall over in ways that make us laugh and heighten our enjoyment in watching them.

The Internet provides an ideal delivery vehicle for these concentrated doses of cuteness. For one thing, videos on the Internet are easily available whenever we want them. In order to have a dish of ice cream, on the other hand, you need to make sure you buy ice cream and stock it in the freezer. Then you need to get up from wherever you are sitting and walk to the freezer. After that, you need to scoop the ice cream into a bowl and then eat it. (Admittedly, Art used to skip the bowl in this last step.) Unlike your desire for cuteness, it requires a bit of work to satisfy your desire for a snack.

Of course, in many circumstances, not only is it difficult to satisfy such a craving, but it may also be inappropriate. You just can't get up in the middle of a meeting or lecture and grab a snack just because you want it (although Bob considers this an appropriate escape mechanism for many meetings). But with the proliferation of Internet-connected devices, you are never very far from a video of cute kittens.

As an added bonus, the kitten videos that go viral are rigged for your pleasure. The people who post these videos take the time to edit the hours of footage they've recorded, and show you only the best parts—the cute expressions or adorable stumbles that make the video maximally enjoyable. The Internet gives you access to *curated* cuteness, almost anytime, almost anywhere.

WITH MOST EXPERIENCES IN LIFE, YOU HAVE TO WAIT FOR THE GOOD PARTS. Bob loves to watch football games. In a typical football game, there are only about eleven minutes of actual action. The rest of the game involves players walking to and from the huddle, time-outs, commercials, and other delays. Of that eleven minutes of action, only about a minute or two involves genuine highlights. That means that Bob routinely spends about three to four hours on a given Saturday or Sunday waiting for 120 seconds of joy.

Now, Bob can justify that time, because football is a social occasion for him. When he and his wife go to University of Texas games, they enjoy the time with each other and their friends and family who sit around them. They can talk, complain about bad calls by the refs, and catch up on one another's lives. And when the exciting moments come, they can cheer.

But nobody would watch a kitten video if it required slogging through an hour of sleeping kitty before the first moment of cuteness. The typical video doesn't require waiting that long. Often, the kitten is doing something wonderful a few seconds after the video starts. And that makes these videos the perfect mental snack break.

Of course, a kitten video can provide a great social experience as well. One reason people share these videos on social networks is so they can also enjoy friends' reactions. Thus, the videos provide a way to share joyful experiences even at great distances.

YOU MIGHT WONDER WHETHER KITTEN VIDEOS, LIKE ANY GREAT LITTLE SNACK, are actually good for you. Because they, like many drugs, can provide an immediate boost of positive feeling whenever you'd like one, you might suspect that they are not something you should waste your time on.

But it turns out that cute videos are probably good for you, as long as you don't become so enthralled that you allow hours to slip by as you sit motionless staring at them. The modern world is filled with lots of stresses. From early in the school years, kids have busy schedules with lots of activities and homework. Most adults' jobs are busy, and many adults work long hours. For many people, days are filled with more drudgery than joy.

There is lots of good evidence that a positive mood is beneficial in lots of ways. For example, people tend to be more creative when they are in a positive mood than when they are in a neutral mood. They make better decisions in complex environments when they are in a positive mood. In addition, people in positive moods have more self-control resources. When you are feeling good, you are less likely to snap at a co-worker who annoys you than when you are not.

And when you are feeling a little down (or perhaps just not feeling that good), a kitten video can jump-start a positive mood. Once you are in a positive mood, two good things can happen. First, people's behavior is often contagious. If you smile, the people around you are likely to smile as well. That means that making yourself a little happier can increase the happiness of the people around you.

Second, your mood affects what you remember. When you feel good, you tend to remember positive things. When you feel bad, you tend to remember sad and stressful things. What you recall in turn influences your mood, so if you start out in a positive mood, that mood can be maintained by the things you remember. When you start out in a bad mood, the things you recall can make the bad mood persist.

If you find yourself stuck in a bad mood, then watching some cute videos can provide a way to inject some positive feeling into your day in ways that might set you on the course to feeling good the rest of the day.

Of course, small doses are the key to any mood enhancement. For one thing, you get the biggest effect on mood from the first video you watch. After that, you have already engaged the system that creates positive feeling from faces. Before long, you may habituate to the videos (as you may habituate to the effects of a drug), meaning that you are no longer getting any additional value from watching them.

Plus, if you spend too much time watching cute videos, you will probably run out of time to take care of important things you need to accomplish, and the stress of missed deadlines can ruin the joy induced by big-eyed kittens.

(BIG EYES +
SMALL FEATURES +
POOR COORDINATION)
× immediate
availability ×
moderation =
perfect healthy
mental snack +
mood boost

Is nostalgia good or bad?

HARD AS THIS MAY BE TO BELIEVE, YOU'RE COMING TO THE END OF *Brain Briefs*. Yup. This is it. After this chapter, the book is over. Your experience of reading it will start to recede into the past. And when it does, the book will start to get better. Of course, the two of us like it a lot already, but when you look back on it, it will probably seem even better to you than it does right now.

Art got to see this in action not long ago, right before one of his kids moved to New York City. The two of them decided to spend the day together having a *Star Wars* marathon. They watched the original trilogy of movies (helpfully numbered Episodes 4, 5, and 6) back-to-back. (Neither of them could stomach watching Episodes 1 through 3 again. There are limits to how much better things can get when looking back on them.) The two of them shared a love of these movies, so they were excited to sit down and start watching. About twenty minutes into Episode 4, they were surprised to see how cheesy the movies actually were. Somehow, their memories of the movies had this incredible halo of affection that masked the significant overacting by the lead characters.

(And before all the *Star Wars* fans start thinking about sending nasty letters, the two of us both recognize what an achievement the movies were and what a tremendous impact they had on all of the movies that came afterward. But they *are* cheesy.)

Of course, one sign that you are getting old is that you start talking about how much better the past was than the present era. Bob sometimes catches himself talking wistfully about how things used to be, knowing full well that he is beginning to sound like a curmudgeon.

Bob's *nostalgia* causes him to feel warmly toward the past, perhaps with a bit of longing for some bygone era. But why is it that people often think that the past was better than the present, even though we know that some aspects of previous decades were downright horrible?

Well, several factors interact to create this sense.

First, as we have discussed previously, we tend to think about things that are psychologically distant from us more abstractly than we think of things that are near. Our past is further from us (in time) than the present, so we think about it more abstractly than the present.

Much of what you may find annoying about the present has to do with specific problems you are currently confronting. Assignments at work or school are a source of stress. You may be worried about something that's broken and needs repair. A co-worker may be trying to talk while you're trying to work. And there are always chores that need to be done that may interfere with your enjoyment of every day.

When you look back on the past, though, many of these specific annoyances are less prominent. Instead, you focus on more general aspects of the past, and many of them are joyous. You remember holiday meals with family, walks in the woods, long drives, and vacations. It is hard to recall in detail all of the petty annoyances you experience day to day, so your fuzzy recollections of the past are often more positive than what you experienced at the time. The trifling arguments with family, itchy mosquito bites after a walk in the woods, boredom during car trips, and stress over finding a gas station when your gauge is close to E all fade from memory when you consider the past from a distance.

A second factor that tends to make people feel wistful is that we think about the past knowing how the story turns out. For example, adolescence is a time that is filled with stress, angst, and frustration. Teenagers are struggling to become independent of their parents, and they are learning to find their place in a social structure. Looking back as an adult on your high school years, though, you know that you survived the difficult moments. You know that many of the events that seemed consequential at the time did not have any lasting impact on your life.

As a result, you are free to focus on the joys of the teen years, like going to high school football games, spending time with friends, or being able to sleep until noon on weekends. It is easy to minimize the stressful moments and focus on the positive elements of the past when you know how the story turned out.

When thinking about the present, you don't know how the story is going to turn out. You don't know whether the problems that cause you stress now will be resolved, or whether they will continue to be problems far into the future. The uncertainty of the present makes it seem more stressful and less enjoyable than the past.

A third factor relates to the *change of standard effect* that we discussed earlier. You often remember general evaluations from the past without remembering the basis of those judgments. That is what probably happened with Art and his son when they went back to watch the *Star Wars* films. They remembered the films as being amazing. But the term *amazing* is a judgment based on a comparison between the movies (when they saw them) and other movies that they could use as a basis of comparison.

Over time, Art and his son saw many more movies, and their definition of *amazing* changed. But their original evaluation of the movies remained in their memories, so when they went back to watch the movies again, they were expecting an amazing experience. Instead, they realized that their beliefs about what counts as amazing had gotten more sophisticated over time. And the original movies now seemed more cheesy than amazing. That said, Art and his son have agreed to continue to think of the movies as amazing despite the disappointing movie marathon.

THIS BRINGS US TO THE QUESTION: IS NOSTALGIA GOOD OR BAD?

There are ways that nostalgia can be bad. If you really believe that the past was so much better than the present, you may lose your motivation to try to make your world a better place.

But nostalgia most often seems to have the opposite effect. In particular, the problems of today can sometimes seem overwhelming. When you look back at the past positively, there are two good things that happen. First, you recognize that you have overcome many problems. Second, you often remember the many people who have helped you through difficult situations in your life. As a result, your positive thoughts about the past actually make you feel more socially connected to your community in ways that make you more optimistic about your ability to handle any challenges you may face right now.

In addition, many of the difficulties of the past were things you couldn't really do anything about. For example, several years ago, Art went to Tunisia for a conference. The trip to get to the conference took thirty-six hours and involved three airplanes, four taxis, one van, and a run-in with the Tunisian police (don't ask). At the time it was pretty stressful. Looking back on it, though, the story is a source of entertainment. There was nothing Art could have done about it anyway, so rather than continuing to think about his trip to Tunisia as a stressful event that he would not repeat, he developed some nostalgia for his African adventure. After all, he did get to spend a week in Tunisia—a pretty unique experience—after the horrendous trek required to get there.

Looking back at uncontrollable events from the past with what some would describe as rose-colored glasses allows people to feel less regret about their life choices, which makes them happier and more self-confident in the present.

And that brings us to the end of our collection of Brain Briefs. We had a great time working together and writing, and we hope you had at least as much fun reading. Thanks for sticking with us. We hope that your later memories of reading this are as positive as our memories of writing it.

DISTANCE MAKES THE *heart* AND *mind* GROW FONDER.

References

1. Does being open to experience lead to success?

Gilovich, T., & Medvec, V. H. (1995). The experience of regret: What, when, and why. *Psychological Review, 102*(2), 379-395.

Kruglanski, A. W., & Webster, D. M. (1996). Motivated closing of the mind: "Seizing" and "freezing." *Psychological Review, 103*(2), 263-283.

Markman, A. (2013). *Habits of leadership.* New York, NY: Perigee Books.

2. Can we really make ourselves happy?

Cacioppo, J. T., Hawkley, L. C., Kalil, A., Hughes, M. E., Waite, L., & Thisted, R. A. (2008). Happiness and the invisible threads of social connection: The Chicago Health, Aging, and Social Relations Study. In M. Eid & R. J. Larsen (Eds.), *The science of subjective well-being* (pp. 195-219). New York, NY: Guilford Press.

Diener, E. (2000). Subjective well-being: The science of happiness and a proposal for a national index. *American Psychologist, 55*(1), 34-43.

Epley, N., & Schroeder, J. (2014). Mistakenly seeking solitude. *Journal of Experimental Psychology: General, 143*(5), 1980-1999.

Fujita, F., & Diener, E. (2005). Life satisfaction set point: Stability and change. *Journal of Personality and Social Psychology, 88*(1), 158-164.

Gilbert, D. T., Pinel, E. C., Wilson, T. D., Blumberg, S. J., & Wheatley, T. P. (1998). Immune neglect: A source of durability bias in affective forecasting. *Journal of Personality and Social Psychology, 75*(3), 617-638.

Maier, S. F., & Seligman, M. E. (1976). Learned helplessness: Theory and evidence. *Journal of Experimental Psychology: General, 105*(1), 3-46.

Seligman, M. E. (2002). *Authentic happiness.* New York, NY: Simon & Schuster.

3. How do we catch a liar?

Ormerod, T. C., & Dando, C. J. (2015). Finding a needle in a haystack: Toward a psychologically informed method for aviation security screening. *Journal of Experimental Psychology: General, 144*(1), 76-84.

Pennebaker, J. W. (2011). *The secret life of pronouns: What our words say about us.* New York, NY: Bloomsbury Press.

ten Brinke, L., Stimson, D., & Carney, D. R. (2014). Some evidence for unconscious lie detection. *Psychological Science, 25*(5), 1098-1105.

4. Should we play brain games?

Baddeley, A. D. (2007). *Working memory, thought, and action.* New York, NY: Oxford University Press.

Newell, A. (1990). *Unified Theories of Cognition.* Cambridge, MA: Harvard University Press.

Newell, A., & Simon, H. A. (1963). GPS: A program that simulates human thought. In E. A. Feigenbaum & J. Feldman (Eds.), *Computers and Thought.* Munich, Germany: R. Oldenbourg KG.

5. Do stories help us remember?

Bransford, J. D., & Johnson, M. K. (1973). Considerations of some problems of comprehension. In W. G. Chase (Ed.), *Visual Information Processing* (pp. 383–438). New York, NY: Academic Press.

Loftus, E. F., & Palmer, J. C. (1974). Reconstruction of automobile destruction: An example of the interaction between language and memory. *Journal of Verbal Learning and Verbal Behavior, 13,* 585–589.

Schank, R. C., & Abelson, R. (1977). *Scripts, plans, goals and understanding.* Hillsdale, NJ: Lawrence Erlbaum Associates.

6. Is pain open to interpretation?

Dewall, C. N., MacDonald, G., Webster, G. D., Masten, C. L., Baumeister, R. F., Powell, C., . . . Eisenberger, N. I. (2010). Acetaminophen reduces social pain: Behavioral and neural evidence. *Psychological Science, 14,* 931–937.

Fernandez, E., & Turk, D. C. (1992). Sensory and affective components of pain: Separation and synthesis. *Psychological Bulletin, 112*(2), 205–217.

Lakoff, G., & Johnson, M. (1980). *Metaphors we live by.* Chicago, IL: University of Chicago Press.

Ramachandran, V. S., Brang, D., & McGeoch, P. D. (2009). Size reduction using mirror visual feedback (MVF) reduces phantom pain. *Neurocase: The neural basis of cognition, 15*(5), 357–360.

Ramachandran, V. S., & Hirstein, W. (1998). The perception of phantom limbs: The D. O. Hebb lecture. *Brain, 12,* 1603–1630.

Wager, T. D., & Atlas, L. Y. (2013). How is pain influenced by cognition? Neuroimaging weighs in. *Perspectives on Psychological Science, 8*(1), 91–97.

7. Do schools teach the way children learn?

Roediger, H. L., & Karpicke, J. D. (2006). The power of testing memory: Basic research and implications for educational practice. *Perspectives on Psychological Science, 1,* 181–210.

Sorce, J. F., Emde, R. N., Campos, J. J., & Klinnert, M. D. (1984). Maternal emotional signaling: Its effect on the visual cliff behavior of 1-year-olds. *Developmental Psychology, 21*(1), 195–200.

Wilson, M. (2002). Six views of embodied cognition. *Psychonomic Bulletin and Review, 9*(4), 625–636.

8. Why do tongue twisters work?

Dell, G. S. (1986). A spreading activation theory of retrieval in sentence production. *Psychological Review, 93*(3), 283–321.

Griffin, Z. M. (2010). Retrieving personal names, referring expressions, and terms of address. *Psychology of Learning and Motivation, 53*, 345–387.

Levelt, W. J. M. (1989). *Speaking: From intention to articulation.* Cambridge, MA: MIT Press.

9. Do we get more done when we multitask?

Altmann, E. M., Trafton, J. G., & Hambrick, D. Z. (2014). Momentary interruptions can derail the train of thought. *Journal of Experimental Psychology: General, 143*(1), 215–226.

Glucksberg, S., & Cowen, G. N. (1970). Memory for nonattended auditory material. *Cognitive Psychology, 1*(2), 149–156.

Pashler, H. E. (1998). *The psychology of attention.* Cambridge, MA: MIT Press.

Schneider, W., & Shiffrin, R. M. (1977). Controlled and automatic human information processing: I. Detection, search, and attention. *Psychological Review, 84*(1), 1–66.

Watson, J. M., & Strayer, D. L. (2010). Supertaskers: Profiles in extraordinary multitasking ability. *Psychonomic Bulletin and Review, 17*(4), 479–485.

10. Can we be conscientious *and* creative?

King, L. A., McKee-Walker, L., & Broyles, S. J. (1996). Creativity and the five-factor model. *Journal of Research in Personality, 30*(2), 189–203.

Wason, P. C., & Johnson-Laird, P. N. (1972). *Psychology of reasoning structure and content.* London, UK: Routledge.

11. Is it true that we only use 10 percent of our brains?

Bear, M. F., Connors, B. W., & Paradiso, M. A. (2015). *Neuroscience: Exploring the brain.* New York, NY: Wolters Kluwer.

12. Is our memory doomed to fail?

Evans, D. A., Beckett, L. A., Albert, M. S., Hebert, L. E., Scherr, P. A., Funkenstein, H. H., & Taylor, J. O. (1993). Level of education and change in cognitive function in a community population of older persons. *Annals of Epidemiology, 3*(1), 71–77.

Hartshorne, J. K., & Germine, L. T. (2015). When does cognitive functioning peak? The asynchronous rise and fall of different cognitive abilities across the life span. *Psychological Science, 26*(4), 433–443.

Salthouse, T. A. (2004). What and when of cognitive aging. *Current Directions in Psychological Science, 13*(4), 140–144.

Thomas, A. K. & Dubois, S. J. (2011). Reducing the burden of stereotype threat eliminates age differences in memory distortion. *Psychological Science, 22*(12), 1515–1517.

13. Why are continuity errors in movies difficult to catch?

Rensink, R. A., O'Regan, J. K., & Clark, J. J. (1997). To see or not to see: The need for attention to perceive changes in scenes. *Psychological Science, 8*(5), 368–373.

Simons, D. J., & Levin, D. T. (1998). Failure to detect changes to people during a real-world interaction. *Psychonomic Bulletin and Review, 5*(4), 644–649.

Zacks, J. (2014). *Flicker: Your brain on movies.* New York, NY: Oxford University Press.

14. Are all narcissists alike?

Carlson, E. N., & Lawless DesJardins, N. (2015). Do mean guys always finish first or just say they do? Narcissists' awareness of their social status and popularity over time. *Personality and Social Psychology Bulletin, 41*(7), 901–917.

Konrath, S., Meier, B. P., & Bushman, B. J. (2014). Development and validation of the single item narcissism scale. *PlosONE, 9*(8), e103469.

Krizan, Z., & Johar, O. (2015). Narcissistic rage revisited. *Journal of Personality and Social Psychology, 108*(5), 784–801.

15. Does time speed up as we get older?

Avni-Babad, D., & Ritov, I. (2003). Routine and the perception of time. *Journal of Experimental Psychology: General, 132*(4), 543–550.

Coane, J. H., & Balota, D. A. (2009). Priming the holiday spirit: Persistent activation due to extraexperimental experiences. *Psychonomic Bulletin and Review, 16*(6), 1124–1128.

Csikszentmihalyi, M. (1990). *Flow.* New York, NY: Harper Perennial

16. Why is forgiveness so powerful?

Freedman, S. R., & Enright, R. D. (1996). Forgiveness as an intervention goal with incest survivors. *Journal of Counseling and Clinical Psychology, 64*(5), 983–992.

Noreen, S., Bierman, R., & MacLeod, M. D. (2014). Forgiving you is hard, but forgetting seems easy: Can forgiveness facilitate forgetting? *Psychological Science, 25*(7), 1295–1302.

Steiner, M., Allemand, M., & McCullough, M. E. (2012). Do agreeableness and neuroticism explain age differences in the tendency to forgive others? *Personality and Social Psychology Bulletin, 38*(4), 441–453.

17. Is our thinking *ever* coherent?

Baddeley, A. D. (2007). *Working memory, thought, and action.* New York, NY: Oxford University Press.

Festinger, L. (1956). *A theory of cognitive dissonance.* Stanford, CA: Stanford University Press.

Higgins, E. T., & Stangor, C. (1988). A "change-of-standard" perspective on the relations among context, judgment and memory. *Journal of Personality and Social Psychology, 54*(2), 181–192.

Read, S. J., Monroe, B. M., Brownstein, A. L., Yang, Y., Chopra, G., & Miller, L. C. (2010). A neural network model of the structure and dynamics of human personality. *Psychological Review, 117*(1), 61–92.

Russo, E. J., Medvec, V. H., & Meloy, M. G. (1996). The distortion of information during decisions. *Organizational Behavior and Human Decision Processes, 66*, 102–110.

Simon, D., & Holyoak, K. J. (2002). Structural dynamics of cognition: From consistency theories to constraint satisfaction. *Personality and Social Psychology Review, 6*(6), 283–294.

Thagard, P. (1989). Explanatory coherence. *Behavioral and Brain Sciences, 12*, 435–502.

Tulving, E. (1983). *Elements of episodic memory.* New York, NY: Oxford University Press.

18. Are our beliefs consistent with one another?

Baron, J., & Spranca, M. (1997). Protected values. *Organizational Behavior and Human Decision Processes, 70*(1), 1–16.

Platt, J. R. (1964). Strong inference. *Science, 146*, 347–352.

19. Why is it hard to learn a new language?

Eimas, P. D. (1971). Speech perception in infants. *Science, 171*, 303–306.

Gomez, R. L., & Gerken, L. (1999). Artificial grammar learning by

1-year-olds leads to specific and abstract knowledge. *Cognition, 70*(2), 109–135.

Newport, E. L. (1990). Maturational constraints on language learning. *Cognitive Science, 14*(1), 11–28.

Saffran, J. R., Aslin, R. N., & Newport, E. L. (1996). Statistical learning by 8-month-old infants. *Science, 274*, 1926–1928.

20. Is our right brain different from our left brain?

Robertson, L. C., & Ivry, R. (2000). Hemispheric asymmetries: Attention to visual and auditory primitives. *Current Directions in Psychological Science, 9*(2), 59–63.

Sperry, R. W. (1974). Lateral specialization in the surgically separated hemispheres. In F. O. Schmitt & F. G. Worden (Eds.), *Neuroscience* (pp. 202–229). Cambridge, MA: MIT Press.

21. How do we overcome writer's block?

Clance, P. R., & Imes, S. (1978). The imposter phenomenon in high achieving women: Dynamics and therapeutic intervention. *Psychotherapy Theory, Research, and Practice, 15*(3), 241–247.

King, S. (2000). *On writing.* New York, NY: Scribner.

Paulus, P. B., Kohn, N. W., & Arditti, L. E. (2011). Effects of quantity and quality instructions on brainstorming. *Journal of Creative Behavior, 45*(1), 38–46.

22. Is failure necessary?

Dunning, D., & Kruger, J. (1999). Unskilled and unaware of it: How difficulties in recognizing one's own incompetence lead to inflated self-assessments. *Journal of Personality and Social Psychology, 77*(6), 1121–1134.

Dweck, C. (2006). *Mindset.* New York, NY: Random House.

Neff, K. (2003). Self-compassion: An alternative conceptualization of a healthy attitude toward oneself. *Self and Identity, 2*(2), 85–101.

Saxenian, A. (1996). *Regional advantage.* Cambridge, MA: Harvard University Press.

23. How much of what we see is real?

Goldmeier, E. (1972). Similarity in visually perceived forms. *Psychological Issues, 8*(1), 1–136.

Kanizsa, G. (1976). Subjective contours. *Scientific American, 234*(4), 48–52.

Palmer, S. E. (1992). Common region: A new principle of perceptual grouping. *Cognitive Psychology, 9*(3), 441–474.

24. Does punishment work?

Higgins, E. T. (1997). Beyond pleasure and pain. *American Psychologist, 52*(12), 1280–1300.

Miller, N. E. (1959). Liberalization of basic S-R concepts: Extensions to conflict behavior, motivation, and social learning. In S. Koch (Ed.), *Psychology: A study of a science. General and systematic formulations, learning, and special processes* (Vol. 2, pp. 196–292). New York, NY: McGraw Hill.

Warr, P. (1999). Well-being and the workplace. In D. Kahneman, E. Diener, & N. Schwarz (Eds.), *Well-being: The foundations of hedonic psychology* (pp. 392–412). New York, NY: Russell Sage Foundation.

25. Why are comparisons so helpful?

Basalla, G. (1988). *The evolution of technology.* Cambridge, UK: Cambridge University Press.

Chen, S., & Andersen, S. M. (1999). Relationships from the past in the present: Significant-other representations and transference in interpersonal life. In M. P. Zanna (Ed.), *Advances in experimental social psychology* (Vol. 31, pp. 123–190). San Diego, CA: Academic Press.

Gentner, D. (1983). Structure-mapping: A theoretical framework for analogy. *Cognitive Science, 7,* 155–170.

Gentner, D., & Markman, A. B. (1997). Structural alignment in analogy and similarity. *American Psychologist, 52*(1), 45–56.

Goswami, U., & Brown, A. L. (1989). Melting chocolate and melting snowmen: Analogical reasoning and causal relations. *Cognition, 35,* 69–95.

Linsey, J. S., Wood, K. L., & Markman, A. B. (2008). Modality and representation in analogy. *Artificial Intelligence for Engineering Design, Analysis, and Manufacturing, 22*(2), 85–100.

Zhang, S., & Markman, A. B. (1998). Overcoming the early entrant advantage: The role of alignable and nonalignable differences. *Journal of Marketing Research, 35,* 413–426.

26. Why do people choke under pressure?

Beilock, S. (2011). *Choke: What the secrets of the brain reveal about getting it right when you have to.* New York, NY: Atria Books.

DeCaro, M. S., Thomas, R. D., Albert, N. B., & Beilock, S. L. (2011). Choking under pressure: Multiple routes to skill failure. *Journal of Experimental Psychology: General, 140*(3), 390–406.

Goetz, T., Bieg, M., Ludtke, O., Pekrun, R., & Hall, N. C. (2013). Do girls really experience more anxiety in mathematics. *Psychological Science, 24*(10), 2079–2087.

Gray, R. (2004). Attending to the execution of a complex sensorimotor skill: Expertise differences, choking, and slumps. *Journal of Experimental Psychology: Applied, 10*(1), 42–54.

Masters, R. S. W. (1992). Knowledge, knerves, and know-how: The role of explicit versus implicit knowledge in the breakdown of a complex motor skill under pressure. *British Journal of Psychology, 83*, 343–358.

Steele, C. M., & Aronson, J. (1995). Stereotype threat and the intellectual test performance of African Americans. *Journal of Personality and Social Psychology, 69*(5), 797–811.

Worthy, D. A., Markman, A. B., & Maddox, W. T. (2009). Choking and excelling at the free throw line. *International Journal of Creativity & Problem Solving, 19*, 53–58.

Worthy, D. A., Markman, A. B., & Maddox, W. T. (2009). Choking and excelling under pressure in experienced classifiers. *Attention, Perception, and Psychophysics, 71*, 924–935.

27. How do we decide what to buy?

Dempsey, M. A., & Mitchell, A. A. (2010). The influence of implicit attitudes on choice when consumers are confronted with conflicting attribute information. *Journal of Consumer Research, 37*, 614–625.

Fader, P. S., & Lattin, J. M. (1993). Accounting for heterogeneity and nonstationarity in a cross-sectional model of consumer purchase behavior. *Marketing Science, 12*(3), 304–317.

Payne, J. W., Bettman, J. R., & Johnson, E. J. (1993). *The adaptive decision maker.* New York, NY: Cambridge University Press.

Simon, H. A. (1957). *Models of man: Social and rational.* New York, NY: Wiley.

Simonson, I. (1989). Choice based on reasons: The case of attraction and compromise effects. *Journal of Consumer Research, 16*, 158–174.

Zajonc, R. B. (1968). Attitudinal effects of mere exposure. *Journal of Personality and Social Psychology, 9*, 1–27.

28. What is the best way to brainstorm?

Finke, R. A., Ward, T. B., & Smith, S. M. (1992). *Creative cognition: Theory, research, and applications.* Cambridge, MA: MIT Press.

Linsey, J. S., Clauss, E. F., Kurtoglu, T., Murphy, J. T., Wood, K. L., & Markman, A. B. (2011). An experimental study of group idea generation techniques: Understanding the roles of idea representation and viewing methods. *Journal of Mechanical Design, 133*(3). doi: 10.1115/1.4003498

Mullen, B., Johnson, C., & Salas, E. (1991). Productivity loss in brainstorming groups: A meta-analytic integration. *Basic and Applied Social Psychology, 12*(1), 3–23.

Osborn, A. (1957). *Applied imagination*. New York, NY: Scribner and Sons.

Paulus, P. B., & Brown, V. R. (2002). Making group brainstorming more effective: Recommendations from an associative memory perspective. *Current Directions in Psychological Science, 11*, 208–212.

29. Why is online communication so ineffective?

Clark, H. H. (1996). *Using language*. New York, NY: Cambridge University Press.

30. Is it possible to remember something that didn't happen?

Johnson, M. K., Hashtroudi, S., & Lindsay, D. S. (1993). Source monitoring. *Psychological Bulletin, 114*(1), 3–28.

Loftus, E. F., & Palmer, J. C. (1974). Reconstruction of automobile destruction: An example of the interaction between language and memory. *Journal of Verbal Learning and Verbal Behavior, 13*, 585–589.

Roediger, H. L., & McDermott, K. B. (1995). Creating false memories: Remembering words not presented in lists. *Journal of Experimental Psychology: Learning, Memory, and Cognition, 21*(4), 803–814.

Thomas, A. K., & Loftus, E. F. (2002). Creating bizarre false memories through imagination. *Memory and Cognition, 30*(3), 423–431.

Wilson, B. M., Mickes, L., Stolarz-Fantino, S., Evrard, M., & Fantino, E. (2015). Increased false-memory susceptibility after mindfulness meditation. *Psychological Science, 26*(10), 1567–1573.

31. Is prejudice avoidable?

Brewer, M. B. (1979). In-group bias in the minimal intergroup situation: A cognitive-motivational analysis. *Psychological Bulletin, 86*(2), 307–324.

Cameron, J. A., Alvarez, J. M., Ruble, D. N., & Fuligni, A. J. (2001). Chidren's lay theories about ingroups and outgroups: Reconceptualizing research on prejudice. *Personality and Social Psychology Review, 2*, 118–128.

Hirschfeld, L. A. (1996). *Race in the making*. Cambridge, MA: MIT Press.

32. What's the best way to deal with life's endless litany of nuisances?

Bushman, B. J., Baumeister, R. F., & Stack, A. D. (1999). Catharsis, aggression, and persuasive influence: Self-fulfilling or self-defeating prophecies? *Journal of Personality and Social Psychology, 76*(3), 367–376.

Findley, M. J., & Cooper, H. M. (1983). Locus of control and academic achievement: A literature review. *Journal of Personality and Social Psychology, 44*(2), 419–427.

Lefcourt, H. M. (1991). Locus of control. In J. P. Robinson, P. R. Shaver, & L. S. Wrightsman (Eds.), *Measures of personality and social psychological attitudes* (Vol. 1, pp. 413–499). San Diego, CA: Academic Press.

Maier, S. F., & Seligman, M. E. (1976). Learned helplessness: Theory and evidence. *Journal of Experimental Psychology: General, 105*(1), 3–46.

33. Is mind reading a necessary skill?

Baron-Cohen, S., Leslie, A. M., & Frith, U. (1985). Does the autistic child have a "theory of mind?" *Cognition, 21*(1), 37–46.

Clark, H. H. (1996). *Using language.* New York, NY: Cambridge University Press.

Ding, X. P., Wellman, H. M., Wang, Y., Fu, G., & Lee, K. (2015). Theory-of-mind training causes honest young children to lie. *Psychological Science, 26*(11), 1812–1821.

Keysar, B. (1994). The illusory transparency of intention: Linguistic perspective-taking in text. *Cognitive Psychology, 26,* 165–208.

Perner, J. (1993). *Understanding the representational mind.* Cambridge, MA: MIT Press.

34. What are brains for, anyway?

Castellucci, V., Pinsker, H., Kupfermann, I., & Kandel, E. R. (1970). Neuronal mechanisms of habituation and dishabituation of the gill-withdrawal reflex in Aplysia. *Science, 167*(3926), 1745–1748.

Clark, A. (2013). Whatever next? Predictive brains, situated agents, and the future of cognitive science. *Behavioral and Brain Sciences, 36*(03), 181–204. http://doi.org/10.1017/S0140525X12000477

Schultz, W., & Dickinson, A. (2000). Neuronal coding of prediction errors. *Annual Review of Neuroscience, 23*(1), 473–500. http://doi.org/10.1146/annurev.neuro.23.1.473

Van Doorn, G., Paton, B., Howell, J., & Hohwy, J. (2015). Attenuated self-tickle sensation even under trajectory perturbation. *Consciousness and Cognition, 36,* 147–153. http://doi.org/10.1016/j.concog.2015.06.016

35. Does listening to Mozart make us smarter?

Chabris, C. F. (1999). Prelude or requiem for the "Mozart Effect"? *Nature, 400*(6747), 826–827. http://doi.org/10.1038/23608

Duke, R. A. (2000). The other Mozart Effect: An open letter to music educators. *Update: Applications of Research in Music Education, 19*(1), 9–16. http://doi.org/10.1177/875512330001900103

Isen, A. M., & Labroo, A. A. (2003). Some ways in which positive affect facilitates decision making and judgment. In S. L. Schneider & J. Shanteau (Eds.), *Emerging perspectives on judgment and decision research* (pp. 365–393). New York, NY: Cambridge University Press.

Rauscher, F. H., Shaw, G. L., & Ky, C. N. (1993). Music and spatial task performance. *Nature, 365*(6447), 611. http://doi.org/10.1038/365611a0

Schellenberg, E. G., & Hallam, S. (2005). Music listening and cognitive abilities in 10- and 11-year-olds: The blur effect. *Annals of the New York Academy of Sciences, 1060*(1), 202–209. http://doi.org/10.1196/annals.1360.013

Steele, K. M. (2006). Unconvincing evidence that rats show a Mozart Effect. *Music Perception: An Interdisciplinary Journal, 23*(5), 455–458. http://doi.org/10.1525/mp.2006.23.5.455

36. Why are other people such slackers?

Christiansen, A., Sullaway, M., & King, C. E. (1983). Systematic error in behavioral reports of dyadic interaction: Egocentric bias and content effects. *Behavioral Assessment, 5*(2), 129–140.

Gilovich, T., Medvec, V. H., & Savitsky, K. (2000). The spotlight effect in social judgment: An egocentric bias in estimates of the salience of one's own actions and appearance. *Journal of Personality and Social Psychology, 78*(2), 211–222.

Kahneman, D., Knetsch, J. L., & Thaler, R. H. (1991). Anomalies: The endowment effect, loss aversion and status quo bias. *Journal of Economic Perspectives, 5*(1), 193–206.

Trope, Y., & Liberman, N. (2003). Temporal construal. *Psychological Review, 110*(3), 403–421.

37. Can delusion be a good thing?

Brehm, J. W., & Self, E. A. (1989). The intensity of motivation. *Annual Review of Psychology, 40*, 109–131.

Chambers, J. R., & Windschitl, P. D. (2004). Biases in social comparative judgments: The role of nonmotivated factors in above-average and comparative-optimism effects. *Psychological Bulletin, 130*(5), 813–838.

Dunning, D., & Kruger, J. (1999). Unskilled and unaware of it: How difficulties in recognizing one's own incompetence lead to inflated self-assessments. *Journal of Personality and Social Psychology, 77*(6), 1121–1134.

Forbes, D. P. (2005). Are some entrepreneurs more overconfident than others? *Journal of Business Venturing, 20*(5), 623-640.

Locke, E. A., & Latham, G. P. (2002). Building a practically useful theory of goal setting and task motivation: A 35-year odyssey. *American Psychologist, 57*(9), 705-717.

Norem, J. K., & Cantor, N. (1986). Defensive pessimism: Harnessing anxiety as motivation. *Journal of Personality and Social Psychology, 51*(6), 1208–1217.

38. Why do we call a dog a "dog"?

Brown, R. (1958). How shall a thing be called? *Psychological Review, 65*(1), 14–21.

Rosch, E., Mervis, C. B., Gray, W. D., Johnson, D. M., & Boyes-Braem, P. (1976). Basic objects in natural categories. *Cognitive Psychology, 8,* 382-439.

Tanaka, J. W., & Taylor, M. (1991). Object categories and expertise: Is the basic level in the eye of the beholder. *Cognitive Psychology, 23,* 457–482.

39. Why do we love kitten videos so much?

Ashby, F. G., Isen, A. M., & Turken, A. U. (1999). A neuropsychological theory of positive affect and its influence on cognition. *Psychological Review, 106*(3), 529–550.

Bower, G. H. (1981). Mood and memory. *American Psychologist, 36*(2), 129-148.

Hildebrandt, K. A., & Fitzgerald, H. E. (1979). Facial feature determinants of perceived infant attractiveness. *Infant behavior and development, 2,* 329–339.

40. Is nostalgia good or bad?

Cheung, W., Wildshutl, T., Sedikides, C., Hepper, E. G., Arndt, J., & Vingerhoets, A. J. J. M. (2013). Back to the future: Nostalgia increases optimism. *Personality and Social Psychology Bulletin, 39*(11), 1484-1496.

Higgins, E. T., & Stangor, C. (1988). A "change-of-standard perspective" on the relations among context, judgment, and memory. *Journal of Personality and Social Psychology, 54*(2), 181–192.

Acknowledgments

THIS BOOK IS THE SPAWN OF A WEEKLY RADIO SHOW CALLED *Two Guys on Your Head*, an unexpectedly and remarkably successful production of KUT Radio in Austin, Texas.

Our expressions of gratitude must begin with the wonderful people of KUT, who helped create, refine, and maintain the pithy little moments of radio that have aired every Friday for well over two years. Special shout-outs are owed to our engineer David Alvarez, who almost silently tolerates our fluid sense of time and our inability to put the microphone in the right place, and to Program Director Hawk Mendenhall, who recognized the potential in our ragged pilot recording three years ago and supported the show from the beginning. We're also grateful to Stewart Vanderwilt, who signs the papers that keep the show on the air, and to the many other folks at KUT who have offered encouragement and assistance, especially Mike Lee, John Burnett, Bob Branson, and Joy Diaz.

Of course, shows like ours don't stay on the air if no one's listening, and we very much appreciate the many people who enjoy the broadcasts and podcasts and tell us that we make them think. There is no higher praise for a couple of devoted teachers.

We could not have gotten this project launched without the help of Giles Anderson, our fantastic agent, whose suggestions led us to think about putting a book together for the show.

The two of us are incredibly lucky to be married to our best friends, Leora and Judith, who graciously put up with our shenanigans with warm acceptance (right?), and unfailing support and encouragement.

Finally, we are endlessly grateful to the real brains behind *Two Guys on Your Head*, the spectacularly audacious, creatively outrageous, and joyfully contagious Rebecca McInroy. We know of no one with a more unbridled curiosity, love of learning, and appreciation for all that life has to offer than our producer, editor, and dear and loving friend. Her combination of creativity, insightful intelligence, and deft use of Pro Tools® make the two of us sound like we know what we're talking about most of the time, week in and week out.

We dedicate this book to her with deep respect, appreciation, and love.

Index

About the Authors

ART MARKMAN IS THE ANNABEL IRION WORSHAM CENTENNIAL Professor of Psychology and Marketing at The University of Texas at Austin and founding director of the Program in the Human Dimensions of Organizations. He has written over 150 research papers on topics including reasoning, decision making, categorization, and motivation. He brings insights from cognitive science to a broader audience through his blogs at *Psychology Today*, *Fast Company*, and *Harvard Business Review* as well as his radio show/podcast *Two Guys on Your Head*. Art is on the scientific advisory boards of the *Dr. Phil* show and *The Dr. Oz Show*. He is the author of several books, including *Smart Thinking*, *Smart Change*, and *Habits of Leadership*.

BOB DUKE IS THE MARLENE AND MORTON MEYERSON CENTENNIAL Professor and Head of Music and Human Learning at The University of Texas at Austin, where he is University Distinguished Teaching Professor, Elizabeth Shatto Massey Distinguished Fellow in Teacher Education, and director of the Center for Music Learning. He is also an advisor to the Psychology of Learning program at the Colburn Conservatory in Los Angeles. His research on human learning and behavior spans multiple disciplines, including motor skill learning, cognitive psychology, and neuroscience. A former studio musician and public school music teacher, he has worked closely with at-risk children, both in the public schools and through the juvenile justice system. His most recent books are *Intelligent Music Teaching: Essays on the Core Principles of Effective Instruction* and *The Habits of Musicianship*, which he co-authored with Jim Byo of Louisiana State University.